THE
SYNCHRONISATION
SHIFT

CONVERGING QUALITY, REGULATORY, AND INNOVATION IN MEDTECH

THE SYNCHRONISATION SHIFT

CONVERGING QUALITY, REGULATORY, AND INNOVATION IN MEDTECH

From Silos to Synchrony. From Compliance
to Capability. From Innovation to Impact.

Suresh Babu Subba Raju

First paperback edition 2025

ISBN 978-1-80541-846-7 (Paperback)
ISBN 978-1-80541-845-0 (eBook)

www.thesynchronisationshift.com

DEDICATION

OM NAMASHIVAYA

To my Naina (Mr. R. Subba Raju) and Amma (Mrs. S. Bhagya Lakshmi)
Your quiet strength and unconditional love continue to guide me. Your silent sacrifices shaped every step I took.

To my Gurus, Mr. C. R. Pandian and Mr. Senthil Kumar
Thank you for opening the door to the world of quality. Your belief became the foundation of this journey.

To my wife, Geetha
My anchor. My strength. My constant.

To my daughter, Samritha
For every moment you gave up so I could finish this book - thank you. Your patience and love are written into every page.

To my family - Sujatha, Dharmaraja, Nivethitha, Harsha, and Aditya
Your support and presence mattered more than you know. You kept life moving when I paused to write.

To my publisher and the entire publishing team - Mr. Patrick & Team
Thank you for believing in this work and helping bring this vision to life with care and craft.

Last but not least, to you, the reader
Thank you. May this book spark clarity, connection, and the courage to lead boldly, build wisely, and synchronise what truly matters.

This isn't just a book.
It is a tribute.
A journey.
A promise.

— *Suresh Babu Subba Raju*

TABLE OF CONTENTS

SYNC OR STUMBLE: MEDTECH'S DEFINING MOMENT

A Note of Thanks

Before anything else, thank you.

Thank you for picking up this book and for giving your time and attention to a subject that is too often seen as a burden rather than a catalyst. In a world overflowing with information, your decision to read this already sets you apart. It shows leadership, the kind that values integration over inertia and real progress over performative compliance.

Whether you lead quality, navigate regulatory landscapes, shape product strategy, engineer solutions, innovate at the edge, or quietly power the work behind the scenes, this book was written with you in mind.

The Moment of Reckoning

Every transformation starts with a reckoning, that point when the cost of standing still becomes greater than the risk of change.

For MedTech, that point has arrived.

AI, digital therapeutics, virtual trials, and connected devices are no longer "the future." They are here, reshaping what patients expect, what providers demand, and what regulators require. Yet too many organisations are still trying to manage quality, regulation, and innovation through fragmented, outdated systems. The results are predictable: friction, duplication, delays, missed opportunities, and in the worst cases, compromised patient outcomes.

This book is not a quick-fix guide, nor is it a manual of standards.

It is a call to lead differently.

It offers a blueprint for synchronisation, reimagining how quality, regulatory, and innovation teams can work not in silos, but as one coherent, living system. A system that empowers leaders and teams alike across the entire lifecycle and culture of an organisation.

It is not the complexity that breaks MedTech. It is misalignment.

Why This Book Exists

Many of us feel like we are running a race that only gets faster.

Innovators stall, waiting for approvals. Compliance teams drown in documents. Data is trapped in silos. Yet the intent is the same for all of us: to deliver safe, effective, life-changing technologies to patients.

So why do we keep misfiring?

Because quality, regulatory, and innovation are too often treated as separate conversations instead of parts of the same story. Because we are still relying on systems built for a pre-digital world to solve digital-first problems.

This book is a reset. It introduces the SHIFT Framework, five interdependent pillars designed to bring true synchronisation:

- **S - Systems Thinking**: See the whole, not just the parts. Link design, risk, clinical, manufacturing, and post-market as one connected system.
- **H - Harmonisation**: Align purpose, process, roles, and metrics so teams pull in the same direction.
- **I - Integration**: Build quality and regulatory into the flow of innovation, upstream and continuous, not as late-stage gates.
- **F - Feedback**: Treat listening as a discipline. Close the loop across complaints, vigilance, and real-world evidence.
- **T - Trust**: Create the culture, governance, and data foundations where convergence thrives – transparent, audit-ready, and patient-centred.

Each chapter brings SHIFT to life with practical models, human-centred design, and real MedTech cases. You won't just read about what is possible. You will see how to build it step by step in your own organisation.

What You Will Discover

This book is designed to take you on a journey from awareness to action, across five sections:

- **The Wake-Up Call**: We start by confronting the core issue: the widening gap between innovation, quality, and regulation. You will see how misalignment fragments teams, fuels silos, and slows progress and why now is the time for the SHIFT.
- **Building the Synchronised Core**: Here you will learn how to design synchronisation into the system itself, through the Integrated Quality Synchronisation Model (iQSM), global harmonisation, agile governance, and the innovation execution engine. Think of this as the operational blueprint for the modern MedTech organisation.
- **Synchronised Excellence in Action**: Where strategy meets execution. We will explore how to build trust through Quality by Design and Human Factors, embed agility into digital lifecycles, foster collaboration across functions, and harness performance intelligence.
- **The Digital Leap**: As devices get smarter and systems more connected, new risks and opportunities appear. We will dive into AI, digital twins, cybersecurity, and real-world data and show how to balance speed with trust in this new landscape.
- **The Culture Catalyst**: Ultimately, transformation doesn't endure because of frameworks alone, but because of culture and leadership. You will see how synchronisation evolves from a project into a mindset, and finally, into the identity of an organisation.

What You Will Take Away

By the time you finish this book, you will have more than ideas; you will have tools you can put to work right away.

You will see the MedTech lifecycle as a connected system, from R&D through post-market. You will gain a shared language that helps quality, regulatory, and innovation teams work as one. You will have practical models like iQSM and the SHIFT Framework, along with playbooks for aligning projects, documents, systems, and governance.

You will understand how to build trust not only with regulators, but also with patients and your own teams. And most importantly, you will carry the conviction to lead differently.

Synchronisation isn't a brake on progress; it is the engine that makes acceleration possible.

An Invitation to Lead the Shift

This book is not the conclusion of a conversation. It is the beginning of a movement toward smarter systems, empowered teams, and a MedTech future defined by speed, safety, and synergy.

The challenges are real. But so is the opportunity.

In a world of accelerating change, those who synchronise will lead.

Let us begin.

THE WAKE-UP CALL – FROM FRICTION TO FOCUS

E very industry faces a defining moment, a point where the structures that once propelled progress begin to fracture under the weight of speed, complexity, and rising expectations. For MedTech, that moment has arrived.

Innovation is accelerating like never before. Artificial intelligence, wearable biosensors, robotic interventions, and personalised therapeutics are reshaping what is possible in diagnosis, treatment, and care delivery. Yet behind the scenes, many organisations are fighting a silent battle, not against technology itself, but against the systems that were never designed to keep pace with this new era.

In Chapter 1: The Convergence Crisis – Why MedTech Can't Afford Misalignment Anymore, we confront this tension head-on. It is not a minor misalignment; it is a systemic gap that is slowing us down, draining resources, and in some cases, putting patients at risk. Teams work in parallel, not in partnership. Innovation advances while regulatory and quality functions play catch-up. What results is not just inefficiency, it is erosion of trust, time, talent, and value.

We examine real-world consequences, from costly recalls to missed opportunities, and ask the question no one wants to ask but everyone feels: Why are we still operating this way?

But the root of the problem goes beyond structure. In **Chapter 2: From Divide to Design – The SHIFT Toward Convergence**, we go deeper. We explore the cultural and behavioural forces that sustain the divide and the friction between agility and assurance, speed and safety, progress and process. We uncover the quiet drivers of disconnection: misaligned incentives, legacy mindsets, and the absence of shared language. But we also introduce a different path, one grounded not in critique, but in design.

This chapter does not just name the pain. It introduces a philosophy and a framework: the SHIFT mindset. With Systems Thinking, Harmonisation, Integration, Feedback, and Trust as its core principles, SHIFT shows how convergence is not only achievable, but necessary. We explore how leading organisations are dismantling silos, reshaping governance, and uniting quality, regulatory, and innovation teams from the very beginning, not just to comply, but to create, accelerate, and endure.

Section I is both an exposition and an invitation. It holds up a mirror to the realities many teams already know but have lacked the words to name. And it offers something rare in transformation journeys: not just insight, but hope. If you have felt the weight of working around broken systems, the fatigue of duplication, or the frustration of misalignment, you are not alone. Across the industry, the call for synchronisation is growing louder and more urgent.

*"Transformation does not begin with a tool,
a policy, or even a model. It begins with
awareness and the courage to respond."*

This is that moment. And this is your wake-up call.

CHAPTER 1

THE CONVERGENCE CRISIS – WHY MEDTECH CAN'T AFFORD MISALIGNMENT ANYMORE

A patient relied on a device. It had passed regulatory review. It had cleared every quality check. And yet, it failed, silently, suddenly, with consequences no checklist could have predicted.

This was not an isolated incident. It was a warning. The message was clear: the MedTech industry faces a convergence problem.

We are at a pivotal moment. Not because of one failure, but because of the widening disconnect between the forces shaping modern MedTech.

On one side, the pace of innovation is breathtaking. Artificial intelligence, software-defined therapies, predictive diagnostics, and connected health ecosystems are no longer "future technologies;" they are redefining what a medical device even is.

On the other side, expectations are rising. Regulatory demands are sharper. Quality standards are higher. Transparency, safety, and global accountability are no longer aspirational, they are non-negotiable. Regulators are watching more closely. Stakeholders are expecting more. Patients are demanding better.

Ideally, innovation delivers breakthroughs, regulation keeps them safe, and quality makes sure they can be trusted again and again. On paper, that is how it should work. In reality, these gears often spin out of sync and operate in isolation, well-intentioned, highly skilled, but unsynchronised. The result is not just inefficiency. It is risk.

What happens when innovation outpaces safety systems? When regulatory strategy becomes an afterthought? When quality becomes a retrospective checklist rather than a proactive discipline? Devices get recalled. Trust erodes. Patients suffer.

This chapter begins with a hard truth: misalignment is no longer a minor process inefficiency, it is a strategic and ethical threat. Fragmentation doesn't just slow progress, it sabotages it. The disconnect between functions is not a friction to manage; it is a failure to address.

And yet, within this crisis lies a powerful opportunity.

Because if we can see the silos, we can dismantle them. If we can acknowledge the misfires, we can design against them. If we can shift from parallel ambition to integrated purpose, we can build systems that are not only compliant, but truly resilient. Not just faster but safer. Not just effective but enduring.

The future of MedTech will not be won by those who excel in isolation. It will be led by those who synchronise.

This chapter is your invitation to begin that journey, not with another tool or template, but with a shift in mindset. From coexistence to convergence, from structural alignment to systemic trust. In practice, this means embedding regulatory, quality, and innovation checks into product strategy from day one, not at the sign-off stage.

The future is already here, and it demands that we move together.

1.1 The Crisis That Should Never Have Happened

1.1.1 A Silent Fault in a Life-Saving Machine

It began quietly, with a recall notice posted without fanfare. A leading infusion pump manufacturer was forced to issue a Class I recall, the most serious category, due to reports of dosing inaccuracies. The pumps had passed all premarket checks. They had been approved for use. Yet patients faced risks of over- or under-infusion, including those in intensive care units.

The issue? A software mismatch. During routine servicing, maintenance teams had inadvertently loaded software intended for one platform onto another. The visual interfaces were similar. The physical hardware appeared compatible. But the internal logic was not. As a result, pumps began to behave unpredictably, delivering incorrect flow rates and displaying confusing information to clinicians. The clinical outcome could be catastrophic, particularly for vulnerable patients requiring precise therapy.

This was not a failure of engineering brilliance, nor of regulatory scrutiny. It was a failure of synchronisation. No one acted maliciously. But checks and controls were not unified across functions. Quality systems lacked post-service validation routines. Regulatory teams were not embedded in servicing protocols. The result: a preventable crisis that impacted hospitals, strained caregivers, and put patients at risk.

1.1.2 How Misalignment Creeps In

To understand how this happened, we must examine the systems that produced it. In many MedTech organisations, innovation, regulation, and quality are treated as separate disciplines. They work to different clocks, speak in jargon the others don't follow, and get rewarded for things that don't always add up to a safe product. While each function pursues excellence within its domain, they rarely share ownership of the product as a whole.

When development teams are praised for speed, but servicing teams are disconnected from quality controls, critical cross-checks are missed. When regulatory compliance is achieved at submission but not revalidated during service events, inconsistencies slip through the cracks.

This convergence deficit is not abstract. It is systemic. It is costly. The takeaway is simple: excellence in isolation always creates risk downstream.

1.1.3 The Hidden Price of Siloed Excellence

Departments often look good on paper: the KPIs are green, the audits are clean, the submissions are on time. But when early insight isn't shared, misalignment grows. Teams learn to optimise in silos only to falter when collaboration is needed most.

The financial cost of a single Class I recall can exceed millions when accounting for remediation, lost revenue, and reputational damage. For staff, recalls often trigger months of firefighting, audits, and burnout, energy that could have gone to innovation. For leadership, the financial shock is equally severe: recalls can wipe hundreds of millions from market value in days. For patients and clinicians, the damage is reputational: trust in both product and company is shaken long after technical fixes are made.

Another case involving software in continuous ventilators illustrates the point. A top-tier manufacturer had to recall life-sustaining ventilators due to a software algorithm that misread battery levels. The ventilator could shut down therapy mid-use, despite having sufficient power. Again, the root cause was a lack of holistic integration across lifecycle stages. Edge-case scenarios had not been tested. Real-world feedback had not influenced design verification. Risk management had not anticipated the exact interplay of patient behaviour and machine logic.

Both these cases were not rooted in engineering incompetence. They were rooted in silos, where quality is reactive, regulation is episodic, and innovation is assumed to be self-contained.

1.1.4 A Pause to Reflect, A Call to Rethink

Each of these failures is also a signal. A wake-up call to rethink how functions collaborate. The way forward is not more red tape. It is deeper alignment. Regulatory, quality, and innovation teams must move beyond coexistence and begin co-creation. That starts with shared visibility, common incentives, and mutual respect across disciplines.

It requires innovation leaders to invite quality and regulatory voices early. It requires regulators to become enablers of safe speed. It demands quality professionals to move beyond audits and become strategic partners in lifecycle design.

Because no matter how brilliant a product is, if it cannot maintain safety through servicing, or if its software is not validated for the real world, the risks will eventually surface, and they will surface in patient outcomes.

1.1.5 When Crisis Becomes Catalyst

This convergence crisis is not a failure of intention. It is a warning. It shows us that excellence within a silo is not enough. The next generation of MedTech demands more than isolated brilliance. It demands connected thinking, continuous feedback, and systems built for synchronisation.

If we respond to this crisis with structural humility and operational courage, we can avoid future recalls, reduce friction, and restore the trust that patients, clinicians, and regulators place in our industry.

This must be the last recall that should not have happened and the first moment we commit to never letting it happen again. The chapters ahead will move from diagnosis to design, showing how regulatory foresight, proactive quality, and integrated innovation can transform crisis into capability.

1.2 Standing at the Crossroads: Change or Cost

1.2.1 A Decade in the Making

There are moments in every industry's evolution when the future presents two distinct paths. For MedTech, that moment has arrived. After decades of remarkable innovation, the industry finds itself grappling with a growing disconnect between speed and safety, creativity and control, vision and verification. The historical separation of innovation, regulation, and quality once allowed for focused expertise and sequential development. But in the current landscape, this separation has become a liability.

Today's medical technologies are no longer static tools. They are learning systems, adaptive platforms, and data-driven engines that evolve long after market release. Artificial intelligence continuously updates algorithms. Software patches alter user interfaces overnight. Cloud connectivity introduces real-time feedback loops. The entire product lifecycle is now dynamic. And yet, many organisations are still operating within static models of governance, documentation, and risk control.

This tension cannot be resolved through isolated improvement. It demands an intentional shift in how we think, design, and lead. Innovation must no longer be treated as the creative sprint that precedes compliance. It must be embedded within a broader strategy that recognises the importance of real-time oversight, proactive quality, and anticipatory regulation. This is not a call for more control, but for smarter coordination.

1.2.2 Technology's Acceleration and the Friction It Creates

Technological acceleration is not a future challenge. It is today's reality. Consider the rise of digital diagnostics, implantable devices with AI-driven response systems, and real-time therapeutic algorithms. These advances are not just iterative improvements. They are paradigm shifts.

Yet many regulatory frameworks were designed for fixed hardware and static functionality. A firmware update that alters clinical response patterns might not trigger a formal regulatory review. A change in user interface that affects patient adherence may go undetected by quality systems focused on specification-level compliance. The result is a growing gap between what the technology does and what the system is prepared to monitor.

This gap creates friction. Development teams are forced to wait for approvals that were not designed for iterative cycles. Quality assurance becomes a bottleneck, trying to enforce controls that were never intended to be applied continuously. Regulatory teams become reactive instead of strategic, constantly playing catch-up with design changes and clinical feedback.

The solution is not to slow innovation, but to rewire the system. Alignment must become the operating principle. When innovation is guided by real-time regulatory insight and continuous quality validation, the speed becomes sustainable. Not risky. Not rushed. But resilient. The actionable insight: build regulatory and quality checkpoints into every iteration, not just the final sign-off.

1.2.3 A Shifting Regulatory Ground

Regulatory authorities around the world are no longer tolerating outdated development behaviours. The shift from MDD to MDR in Europe, and the introduction of more rigorous post-market surveillance and clinical evaluation requirements, signals a new phase of accountability. What used to be sufficient at product launch is now just the beginning of a long and visible journey.

This new landscape requires new thinking. Design inputs must now anticipate long-term evidence generation. Risk files must be living documents, updated through real-world feedback. Traceability must extend not just through development, but across service, update, and patient experience. Regulators are no longer looking only for correctness. They are seeking continuity, transparency, and adaptability.

Organisations that respond well to these changes are those that reimagine the role of regulation from a hurdle to a strategic lens. Regulatory teams must be embedded in product strategy, not just submission planning. They must help shape architectures that scale, not just dossiers that comply. This requires a cultural shift:

"Regulation is no longer the gate at the end of the road. It is the map that guides every turn."

1.2.4 The Rise of Informed and Impatient Stakeholders

The pressure for convergence is not only coming from within the industry. It is being driven by patients, clinicians, payers, and policymakers. Stakeholders are no longer passive. Patients demand access to performance data. Clinicians want evidence of usability and clinical value. Payers require proof of economic benefit. And the public expects transparency, not just in performance, but in purpose.

In this connected ecosystem, a recall or warning letter does not stay confined to regulatory circles. It becomes public knowledge. Trust is not built through compliance statements. It is earned through behaviour, decisions, and responsiveness. Every gap is visible. Every misalignment is magnified.

This shift means that excellence can no longer be defined by what happens behind closed doors. It must be experienced by all stakeholders as cohesion, clarity, and confidence. A product that performs well but is surrounded by process confusion or safety concerns will not earn market trust. And a system that reacts only after failure will not survive increasing public and professional scrutiny.

1.2.5 Leaving the Old Path Behind

The old path, where each function manages its domain in isolation, joining hands only at sign-off, is no longer sustainable. It cannot keep pace with technology. It cannot manage complexity. And it cannot earn the trust of a new generation of patients and professionals.

To thrive, organisations must embrace synchronisation not as a project, but as a principle. Quality must be designed in, not added on. Regulatory foresight must shape innovation, not follow it. And innovation must serve not only functionality, but transparency, responsibility, and real-world relevance.

The winners won't just be the biggest or the fastest. They will be the ones whose teams finally stop pulling against each other and start rowing in the same direction. Their teams will speak with one voice. Their systems will update in harmony. Their decisions will be anticipatory, not reactive.

This is not about abandoning the structure. It is about redefining it. This is not about weakening rigour. It is about distributing it. And most importantly, this is not about choosing between innovation, regulation, and quality. It is about recognising that only together can they deliver what MedTech now demands: sustained impact and uncompromised trust.

1.3 When Silos Become Sinkholes

1.3.1 Lost in Translation Inside One Company

Every day in MedTech organisations, a silent tension plays out. Research and development teams focus on technical feasibility and speed to market. Regulatory teams interpret evolving requirements and plan for global approvals. Quality assurance teams work to ensure process discipline, risk mitigation, and traceability. All of them sit in the same buildings, attend the same all-hands meetings, and aim to serve the same patients. Yet, they operate as if separated by borders.

The issue isn't lack of commitment or capability. It is that each function speaks a fundamentally different language. Engineers speak in lines of code and cycles of iteration. Regulatory professionals converse in standards, submissions, and risk classifications. Quality teams engage in CAPAs, validations, and audits. These vocabularies are individually rich but rarely translated. Cross-functional meetings become decoding exercises rather than decision-making platforms.

This linguistic misalignment breeds frustration. Regulatory is seen as a barrier. Quality is labelled as delay. Innovation is misunderstood as reckless speed. The truth is none of these teams are wrong. But none of them are aligned. And when misalignment becomes systemic, the risks begin to grow, unseen, unchallenged, and, eventually, uncontrollable.

1.3.2 The Consequences of Compartmentalised Brilliance

Organisational silos were once a strength. They allowed expertise to deepen and accountability to sharpen. But in today's MedTech reality, where every product is a blend of hardware, software, and data, this compartmentalisation is no longer sustainable.

When each function is rewarded in isolation - speed for development, completeness for regulatory, zero findings for quality - the result is fragmented success. The product may ship, the audit may pass, the approval may come through, but those wins can mask deeper systemic fragilities.

Miscommunication between teams leads to late discovery of gaps. A missed update in the design file can trigger a nonconformance in audit. A requirement changes understood by engineering but not captured by regulatory can delay submission. A usability concern flagged by quality but dismissed as low-risk may emerge as a post-market complaint pattern.

These are not fictional hypotheticals. They are recurring realities that surface too late, too often. The true damage lies not only in the failure itself, but in the delayed recognition that the system never caught it. Because the system never acted as one.

1.3.3 When Alignment Is An Afterthought

In many organisations, cross-functional alignment is treated as a milestone, a box to check at phase gates or design reviews. It is something to achieve, rather than something to embody. This reactive mindset turns collaboration into a rescue operation instead of a shared design principle.

Decisions made in isolation are rarely neutral. A shortcut in design can cascade into regulatory ambiguity. A change in classification can unravel a validated process. A clinical insight missed in the development phase can lead to years of post-market struggle. These issues are not inevitable. They are consequences of delayed integration.

The most critical insight is that alignment that happens late is not alignment; rather it is remediation. And remediation, even when well-executed, is a signal that the system failed upstream. If teams only come together at the end, that is not alignment. That is a last-minute scramble and by then, something critical has usually already slipped through. The lesson: real alignment begins on day one, not at the final review.

1.3.4 Reimagining Collaboration at the Core

The path forward demands a cultural shift, not just a procedural one. Synchronisation must begin where ideas begin, in design meetings, in requirement discussions, and in early trade-off debates. Regulatory professionals must not be consulted after design freeze but involved at architecture definition. Quality experts must not be invited to verify after the fact but empowered to influence before the first prototype is built.

Collaboration, in its highest form, is not a compromise. It is co-creation. When all functions share accountability for both product and process, when they shape decisions together, the result is stronger than any individual contribution. This is where convergence truly begins, not in the merging of documents, but in the merging of intent.

Successful organisations make this visible. Their roadmaps are built with shared ownership. Their design reviews include voices from every function. Their incentive structures reward collective impact, not just departmental output. They measure success not only by how fast they move, but by how few surprises they encounter downstream.

What is needed is not more communication, but more connection. The kind that transforms a regulatory checklist into a product enabler. The kind that turns quality feedback into design foresight. The kind that allows innovation to flourish without fear of hidden landmines.

Silos are no longer tenable. Not because they are evil, but because they are insufficient. Excellence in isolation is no longer enough. What matters now is excellence in alignment: teams thinking together, planning together, and executing with the shared rhythm of synchronised intent.

1.4 The Real Cost of Misalignment

1.4.1 When Gaps Turn Into Failures

Misalignment in MedTech rarely begins as a dramatic failure. It begins subtly as an undocumented design decision, a delayed risk review, or a missed clarification between regulatory and development. These gaps are small at first. Easy to ignore. Until they aren't.

Over time, these small disconnects become patterns. A design update is implemented without regulatory input, assuming no impact. A risk file remains static, even as product functionality evolves. A usability finding is logged but never escalated because timelines are tight. The system continues to function. Products are launched. Inspections are passed. Confidence remains, until the product fails.

This is the deceptive nature of misalignment. The cracks are invisible until they reach the patient. Then they become real, painful, and often irreversible. What seemed like isolated oversights, missed feedback loops, non-updated documents, and disengaged stakeholders compound into systemic weakness.

These failures are not just operational breakdowns. They are reflections of cultural and structural gaps. And the consequences are rarely confined to one department. When misalignment leads to field action, the entire organisation pays the price.

1.4.2 The Heavy Price of Fragmented Wins

The cost of misalignment cannot be measured in timelines alone. It must be measured in impact: on patients, on trust, and on organisational potential.

From a patient's perspective, these failures are deeply personal. A device that underperforms, a therapy that becomes unavailable, a treatment that is interrupted – these are not regulatory issues, they are life-altering events. Behind every delayed corrective action, there is a human waiting for reliability, reassurance, and care.

From a regulatory standpoint, misalignment can escalate quickly. An overlooked classification shift, an unvalidated software change, or an incomplete post-market file can lead to audit findings, warning letters, market withdrawals, or consent decrees. These events are often traced back not to malice or incompetence, but to disconnection.

And then there is the reputational cost. A single recall can undo years of market trust. Regulators lose confidence. Investors hesitate. Healthcare providers ask harder questions. Employees become disillusioned. When quality, innovation, and regulation are misaligned, no part of the business remains untouched.

More insidiously, misalignment quietly erodes potential. Ideas are stalled by unnecessary delays. Designs are constrained by late-stage rework. Market opportunities are missed because global requirements weren't built in early. Innovation suffers not from lack of talent or intent, but from lack of integration.

1.4.3 Speed Without Alignment, Quality Without Agility

There is a common misconception that speed and safety are opposing forces. That agility must compromise control. But this is a false trade-off. The real challenge is not balancing speed against quality or innovation against regulation. The challenge is achieving synchrony.

When innovation, regulatory, and quality teams move in concert, speed becomes strategic. Fast decisions are possible because risks are assessed in real time. Agile iterations are supported by flexible but robust processes. Submissions become smoother because regulatory expectations have been shaped by design choices made with foresight.

In contrast, quality without innovation becomes defensive. It polices the past instead of enabling the future. Regulation without adaptability becomes a constraint, instead of a catalyst. Innovation without integration risks non-compliance, rework, or even patient harm.

The most resilient MedTech companies do not move slower. They move smarter. They embed regulatory clarity into product design. They build quality into development practices. They test market readiness alongside technical feasibility. In doing so, they achieve the holy grail, velocity with stability.

1.4.4 Rebuilding Coherence, Restoring Trust

Misalignment is not inevitable. It is a symptom of legacy models: linear, siloed, and reactive. Reclaiming coherence requires a systemic reset. Not a new checklist, but a new mindset.

It starts with shared ownership. Cross-functional collaboration must be designed into governance, not added as an afterthought. Objectives must align across functions, so that what drives innovation also strengthens compliance. Metrics must reward not just outputs, but integration. Conversations must shift from tolerance to trust.

Technology can help, but only if it is embedded in a culture of transparency. Traceability platforms, connected documentation, and real-time dashboards are only as powerful as the intent behind them. When used to bridge gaps rather than highlight them, they become tools of cohesion.

But most of all, trust requires leadership. Not just from the top, but from every corner of the organisation. Leaders who challenge functional blind spots. Leaders who promote early collaboration, not just final review. Leaders who measure success by outcomes, not only activities.

This is what restores trust, not only with regulators and customers, but with teams themselves. When functions align, confidence grows. When systems speak the same language, errors are caught earlier. When people believe their contributions are interconnected, they move with greater purpose and conviction.

The cost of misalignment is high. But the reward for convergence, real convergence, is higher. It is not just operational efficiency. It is reputational strength. Patient safety. And strategic freedom.

1.5 Why This Moment Matters

1.5.1 The Alignment Imperative – Why Innovation, Quality, and Regulation Can No Longer Work Apart

There are moments in every industry's evolution when a subtle shift becomes an undeniable imperative. For MedTech, this moment has arrived. It has not come in the form of sudden disruption, but rather as the weight of accelerating complexity pressing against outdated systems. The very structures once built to ensure control, compliance, and coordination now struggle to manage the pace and nature of modern innovation.

For too long, many organisations have operated under a functional truce: innovation to build, quality to verify, and regulation to approve. By necessity, these disciplines are interconnected, yet too often they are treated as separate silos, touching only at milestone gates and reactive reviews. That model can no longer hold.

The products we design today are no longer fixed outputs; they are learning systems, adaptive platforms, and digitally connected solutions. The regulatory bar continues to rise, demanding greater transparency, traceability, and continuous oversight. Quality has evolved beyond an operational function and is emerging as a strategic differentiator. And patients, empowered by information and expectations, demand more than functionality. They seek trust.

In this new landscape, alignment is no longer a luxury. It is a strategic necessity. The capacity to innovate is now directly tied to the ability to integrate. Those who persist with old models, sequential handoffs, late-stage convergence, and isolated metrics will find themselves lagging in performance, trust, and access. The next generation of success will not be driven by the most inventive, but by the most synchronised.

The critical question facing every MedTech leader is no longer just, "Are we doing things, right?" It is, "Are we doing the right things, together?"

1.5.2 An Invitation to Reinvent

This chapter began with a crisis, a recall that revealed not regulatory negligence or design failure, but a missed opportunity for deeper integration. That crisis is not isolated. It is echoed across countless projects, portfolios, and companies. But within that signal lies a profound invitation.

This is not merely a call to rethink tools or restructure teams. It is a chance to reinvent how we work. Not to abandon excellence within functions, but to elevate excellence across them. Not to add complexity, but to reduce friction. Not to blur roles, but to align purpose.

The iQSM model, introduced in the following chapters, builds upon this foundational insight. It recognises that systems must no longer be built for compliance alone, but for adaptability. That regulation must not be retrofitted but embedded. That quality must not slow innovation but strengthen it. And most critically, that leadership must not simply coordinate but connect.

This is not just a call to reflect. It is a call to act. Ask yourself and your teams bold, practical questions like:

- Are regulatory and quality voices shaping decisions at the idea stage?

- Do your systems enable real-time visibility and cross-functional learning?
- Are you building a culture of shared accountability or functional competition?
- Can your processes adapt not just to new markets, but to new models?

The answers will determine not just audit readiness or launch success, but long-term resilience and market leadership.

1.5.3 Toward the Other Side of Crisis

The convergence crisis is not a temporary detour; it is the fault line beneath our current structures. But recognising it gives us the power to reshape what comes next. Not through fear, but through foresight.

Those who succeed in the end are not the ones who avoided failure, but the ones who learned from it. They will have closed the distance between strategy and execution, design and compliance, speed and safety. They will have seen that excellence is not found in isolated metrics but in collective outcomes.

This is not about making compliance invisible. It is about making excellence visible. When integration becomes the standard, when quality is the starting point, not the reaction, everything changes. Products are more reliable. Teams move faster. Trust is easier to earn and harder to lose.

Let this not be another chapter in a history of missed signals. Let it be a turning point. A commitment to act. A choice to synchronise not tomorrow, but today.

Because the next innovation will not wait. The next risk will not signal in advance. And the next patient will not care who was at fault, only whether we got it right.

Let us begin by getting it right, together.

The next chapter invites us to step into that space, not with blame or nostalgia, but with clarity and intention. It explores how cultural friction, process inertia, and structural misfires have kept innovation, quality, and regulatory disciplines running in parallel rather than in partnership. And more importantly, it shows how we can redesign the system, not just to work better, but to work together.

Because the future of MedTech will not be shaped by isolated excellence. It will be built by convergence – thoughtful, synchronised, and sustained. And the journey from divide to design begins now.

CHAPTER 2

FROM DIVIDE TO DESIGN – THE SHIFT TOWARD CONVERGENCE

The MedTech industry stands at a crossroads, not from a lack of innovation, but from a failure to integrate. On one side lies a surge of possibility: intelligent devices, connected care systems, and software-driven diagnostics. On the other lies the drag of outdated processes, fragmented functions, and entrenched silos. This divide, though invisible in dashboards and status reports, has become one of the greatest barriers to meaningful progress.

We are now in an era where medical technologies have outgrown the systems meant to govern them. Development is agile, but compliance remains static. Devices are adaptive, but documentation expects linearity. Innovation is celebrated, yet integration is rarely prioritised. The result is not chaos, but inertia masked as order. Approvals are granted. Launches proceed. But the full promise of breakthrough technology is constrained, not by technical limitations, but by organisational ones.

This chapter is not about patching symptoms, rather it is about root causes. Why do our smartest teams struggle to collaborate across functions? Why does speed often undermine structure, or structure suffocate speed? And what would it take to build an organisation where innovation, regulation, and quality don't merely coexist, but reinforce one another?

We will explore the silent tensions shaping decisions, the cultural habits that reward avoidance over accountability, and the structural designs that isolate instead of integrating. More importantly, we will offer a new path: a SHIFT in mindset, structure, and leadership that enables MedTech organisations to move in rhythm, not rivalry.

Because the future of health technology will not be defined by how fast we invent, but by how well we converge. This chapter marks the turning point, from recognising the divide to redesigning the system. From parallel excellence to purposeful synchrony. From divergence to design.

2.1 Innovation Leaps, Integration Lags

2.1.1 From Bone Saw to Biosensor - The Evolution of Possibility

The history of medical devices is one of remarkable transformation. A generation ago, the pinnacle of surgical precision was a bone saw, a symbol of strength, control, and tactile mastery. Today, that symbol has been replaced by implantable biosensors, autonomous surgical robotics, and AI-driven diagnostics. Devices are no longer passive tools; they are dynamic systems that learn, adapt, and interact.

Cloud-connected monitors, software-updatable implants, and decision-support algorithms are redefining what is possible in diagnosis, treatment, and patient engagement. The shift from mechanical to digital, from reactive to predictive, marks a new frontier in care.

But while technology races forward, the operational systems meant to ensure safety and compliance lag dangerously behind. The chasm between product innovation and organisational integration widens with every breakthrough.

2.1.2 The Innovation-Integration Divide

This divide is not theoretical. Rather, it is felt daily across the product lifecycle. Developers sprint with agile tools and rapid iteration, while regulatory and quality systems, built for static, hardware-dominant products, struggle to keep pace. Agile sprints collide with rigid submission timelines. Continuous updates trigger revalidation requirements never designed for dynamic ecosystems.

This friction is not just inconvenient. It is dangerous. Miscommunication, missed signals, or gaps in oversight create real-world risk: recalls, delays, reputational damage. Worse, products that are technologically advanced but organisationally unsupported may underperform, incur recalls, or erode patient trust.

In this environment, brilliance alone is not enough. What is required is synchronisation, the ability to move as one, from ideation to market, with coherence and shared intent.

> *"Misalignment is not a process hiccup,*
> *but it is an enterprise risk with financial,*
> *reputational, and human costs."*

2.1.3 When Brilliance Meets Bureaucracy

The tension between innovation and governance often plays out in quiet dysfunction. Engineers sprint ahead with prototypes, only to hit slow, opaque approval gates. Quality professionals flag risks early but are excluded from initial design conversations. Regulatory teams prepare for submissions, only to be blindsided by late-stage design changes that rewrite their timelines overnight.

Consider a familiar scene: a device is nearly ready for market, celebrated at an internal demo, when a single overlooked regulatory requirement surfaces. What follows is not just a two-month delay, but millions in added cost, exhausted teams, a frustrated leadership update, and the quiet erosion of trust between functions. What began as brilliance now feels like bureaucracy at its most punishing.

This is not incompetence, rather it is cultural and structural. Teams speak different languages, operate to different rhythms, and are incentivised in isolation. The mission is shared, but the motion is not. And when motion fragments, innovation stumbles, costing not just time, but credibility, morale, and in some cases, patient trust.

> *"Brilliance meets bureaucracy, not by necessity, but by design. And that design is overdue for rethinking."*

2.1.4 Toward a New Model of Harmony

Closing the innovation–integration gap requires more than better tools or updated policies. It demands a shift in how organisations see themselves, not as a series of discrete functions, but as a unified system with shared accountability.

Quality and regulatory perspectives must be embedded not after design, but from its inception. Teams must be equipped to understand one another's constraints, speak a common operational language, and operate with transparency. Integration is not the enemy of speed; it is its safeguard.

The strongest MedTech organisations are not those with the flashiest products, but those whose systems are as innovative as their technology. Integration is not a barrier, it is the bridge.

Because true leadership in this industry will not come from inventing the next great device, but from building the systems that ensure it reaches the patient safely, sustainably, and at scale.

2.2 The Progress–Process Gap

2.2.1 Different Worlds, One Mission

In a sleek innovation lab in Zurich, a design team is testing its third prototype of a wearable AI-based respiratory monitor. Their pace is electric, user feedback is integrated in near real time, software updates are coded weekly, and every iteration pushes closer to a breakthrough.

Meanwhile, in the same building but a different world, the quality and regulatory team prepares documentation for a pre-submission meeting. Their process is precise, disciplined, calibrated for control, not speed.

Both teams are excellent. Both are essential. Yet too often, they operate at odds.

One defines progress as speed of problem-solving. The other defines process as depth of risk control. Between them lies a growing rift, not from intent, but from structure, incentives, and the inability of legacy frameworks to keep up with modern product realities.

"Excellence in isolation is still misalignment - mission success requires shared rhythm, not just shared intent."

2.2.2 Where Progress Collides with Process

The MedTech industry is being reshaped by continuously evolving technologies: cloud updates, machine learning refinements, algorithmic decision loops. Yet the processes supporting them remain anchored in waterfall-era assumptions: fixed baselines, staged reviews, monolithic validations.

For agile teams, iteration feels natural. But even a minor software tweak can trigger major regulatory implications: reclassification, expanded testing, or new post-market reporting. What one team calls "a small fix," another may classify as "a significant change." These dissonances are not theoretical. They create delays, compliance risks, and operational strain.

Change-control systems groan under constant updates. Verification plans fall behind shifting code. Review boards meet after the fact, not alongside development. And too many organisations have normalised this dysfunction.

The result? A vicious cycle of fire-fighting, frustration, and missed opportunity, not because anyone is failing, but because the system is misaligned with the speed of innovation.

2.2.3 Process Without Progress, Progress Without Anchors

When unaddressed, the gap between progress and process pushes organisations into two traps:

1. Caution dominates. Teams avoid improvements because the process to implement them is too heavy.
2. Speed dominates. Agile teams release updates without fully integrating safety, usability, or regulatory foresight.

Both scenarios carry high costs: lost opportunities, increased recalls, brand erosion, regulatory warnings, and staff burnout.

True maturity lies not in choosing between process or progress, but in building systems where they reinforce one another. Process ensures progress is sustainable. Progress ensures process remains relevant.

> *"Balance is not compromised; it is the discipline that makes both speed and safety possible."*

2.2.4 Finding a New Rhythm of Collaboration

Bridging this gap isn't about adding more forms to fill out. It is about changing how people talk to each other. Regulatory and quality can't show up at the end with red pens, they need to be in the room from the first sketch, helping shape decisions before they harden. Engineers, in turn, need to know that every "small" design choice – a software tweak, a material substitution – echoes downstream in safety files, submissions, and ultimately, patient outcomes.

Collaboration works when it is a habit, not a checkpoint. When conversations happen daily instead of in emergency meetings. When the tools let everyone see the same version of the truth in real time, rather than discovering gaps during an audit. And when success is measured not by each team's individual KPIs, but by whether the patient gets a safe, effective device on time.

In my own experience, I have seen what happens when quality is brought in only at the end. As I often remind colleagues:

> *"If you only call me when something*
> *goes wrong, it is already too late."*

That is the mindset shift, from reactive consultation to proactive co-creation.

In that rhythm, compliance doesn't slow you down, it clears the path. Innovation doesn't clash with assurance, rather it makes assurance stronger. Progress and process stop pulling in opposite directions. Instead, they move together, like instruments in the same piece of music, different notes but one performance.

> *"Collaboration becomes powerful when it shifts from*
> *episodic to embedded, from box-ticking to co-creating."*

2.3 Culture Beneath the Surface

2.3.1 Beyond the Surface of Compliance

At a cross-functional meeting in a global MedTech firm, a quality engineer raises a firmware concern in an insulin delivery device. The issue is minor, but under rare conditions could misreport dosage. The team listens politely. The concern is noted. But deadlines loom, resources are stretched, and reopening review feels disruptive. The issue is quietly deprioritised.

I have been in rooms like this. You can almost feel the hesitation in the air. No one argues with the risk, but no one wants to own the delay either. The engineer leaves wondering if they pushed too hard. The project team convinces itself it was "edge-case noise." The meeting moves on.

This is not negligence. It is culture. In many environments, raising concerns feels risky. Compliance is championed in policy, but psychological safety is absent in practice. People begin to self-censor, soften language, or wait for someone else to speak up. Truth gets filtered for "political safety." Over time, the system grows brittle, looking robust on the surface but fragile underneath.

And here is the paradox: the same organisations that preach "patient safety first" often create environments where the very people protecting safety feel least safe to speak. That gap, between what is said and what is tolerated, is where blind spots live.

"A culture without psychological safety guarantees
blind spots, no matter how strong the process appears."

2.3.2 The Weight of the Invisible

Organisational culture is like an iceberg: above the surface are policies, procedures, and stated values. Below lies what truly drives behaviour: unspoken rules, power dynamics, and the emotional cost of dissent. In quality- and regulation-heavy environments, where the stakes are high and scrutiny is relentless, a culture of fear can quietly form.

It manifests in subtle ways. Risk registers are sanitised. Metrics are gamed. Teams avoid raising ambiguity for fear of slowing projects or inviting audits. Leaders may unintentionally reward conformity over curiosity. And silos become reinforced not by org charts, but by fear: fear of blame, fear of rework, fear of not being 'right.'

This is not simply a human resources issue. It is a strategic risk. Because in systems where fear outweighs trust, latent issues remain hidden until they erupt, often in the form of recalls, warning letters, or patient harm.

> *"Fear corrodes compliance, but*
> *only trust can sustain it."*

2.3.3 The Language of Resistance

In most companies, resistance doesn't sound like resistance. It sounds like logic. "This isn't the right time." "We have always done it this way." "Let us wait until the next phase." I have heard these lines from smart, committed people, not because they are lazy, but because they are tired, sceptical, or convinced raising a new idea won't matter.

Resistance is rarely about the change itself. It is about what the change represents: uncertainty, accountability, exposure.

And here is the truth: you don't melt cultural icebergs with another strategy document. You melt them in conversations, in the way leaders model behaviour, in the stories people tell when the meeting ends. That means leaders must ask harder questions, not just "How does our system work?" but "What does it feel like to work in this system?"

2.3.4 From Caution to Courage

Changing culture takes more than posters about "quality first" or slide decks about collaboration. It takes leaders and teams creating spaces where it feels safe to speak up and where raising a hand is seen as protecting patients, not slowing projects.

Picture this: an engineer delays a launch because she spots a firmware risk. Instead of being punished, she is thanked. In that room, curiosity beats defensiveness. Safety isn't "QA's job" anymore, rather it is everyone's job. That is the shift, from caution to courage, from silence to signal.

And it always starts with leadership, with leaders who admit where things went wrong, who invite challenge early, and who back up the people who raise uncomfortable truths. Culture doesn't spread from policy manuals. It spreads from people watching how their leaders act, every day.

Because culture is not what is written on the walls, it is what we tolerate in the halls. It shows up in the side-conversations, in the meetings after the meetings, in how teams behave when the pressure is highest. That is when systems either bend or break.

So instead of just auditing our processes, we should also be auditing our assumptions. Are we rewarding candour? Are we making it safe to tell the truth? If we don't, the iceberg beneath the surface will eventually crack the hull. But if we bring it into the open, if we reshape it together, and it becomes the very foundation we can build on.

2.4 The SHIFT Mindset

2.4.1 From Fragmentation to Fluidity

At the heart of MedTech's transformation lies a fundamental question: can we continue to navigate exponential complexity with linear thinking? The evidence suggests we cannot. Traditional organisational models – sequential, functionally siloed, rigid in structure – were built for a different time. They served us in the era of waterfall development and discrete hardware. But in a world of adaptive software, AI-powered diagnostics, and real-time health ecosystems, these models no longer serve innovation, nor do they serve patients.

The SHIFT mindset offers a new compass. It is not a framework we apply after strategy, but it is the foundation of how we build strategy. It is not about reorganising departments. It is about rewiring how we think, decide, and collaborate across disciplines. This is not incremental improvement. This is an intentional mental pivot from fragmentation to flow, from compliance to contribution, from isolation to integration.

2.4.2 The Five Pillars of SHIFT

The SHIFT mindset rests on five principles that don't just sound good on paper; they make the difference between a system that creaks under pressure and one that adapts with confidence.

S - Systems Thinking

Too often we treat issues as one-off events: a recall, a missed deadline, a CAPA. But nothing in MedTech is truly isolated. A small design tweak in software can ripple into manufacturing, risk files, servicing protocols, and ultimately patient safety. Systems thinking is about seeing those ripple effects early and acting with the whole picture in mind. It is a shift from "my department did its part" to "did the system work end to end?"

H - Harmonisation

Harmonisation doesn't mean making everything identical. It means making it coherent. A global company can't afford to run 10 different compliance frameworks for 10 different markets. That only creates duplication and confusion. Harmonisation means teams share a backbone of processes and language, while still adapting to local realities. The result isn't sameness, it is rhythm.

I - Integration

The biggest delays don't come from technology; they come from people working in sequence instead of together. Integration means quality and regulatory aren't bolted on at the end, but present from day one: in the brainstorm, in the trade-off debates, in the early prototypes. Products built this way don't need to be retrofitted for compliance, rather they are designed to be safe and compliant from the start. That is what real resilience looks like.

F - Feedback

Most companies collect feedback. Too few use it well. Complaints pile up, audit findings get filed away, and post-market data arrives too late to shape the next release. In a SHIFT culture, feedback is treated like oxygen: constant, vital, non-negotiable. Engineers hear from clinicians. Designers see usability data. Leaders act on patient stories, not just KPIs. Feedback isn't noise, it is the signal that keeps the system honest.

T - Trust

Trust is the invisible infrastructure holding it all together. Without it, teams second-guess each other, hide problems, and delay decisions. With it, people speak up early, share uncertainty, and take ownership. Trust means regulatory is treated as a partner, not an obstacle. It means leaders model candour by admitting what they don't know. And it means people believe the system has their back when they raise a concern. Trust isn't a policy. It is a daily practice.

2.4.3 Beyond Compliance, Toward Contribution

In many companies, regulatory and quality are still seen as the brakes on innovation: important, yes, but mostly reactive. The SHIFT mindset flips that script. It asks us to see these functions not as the "police," but as partners. Not the people who slow things down, but the people who make sure the speed is safe, sustainable, and trusted.

This isn't just a matter of language. When regulatory and quality are positioned as contributors from the very beginning, the entire rhythm of development changes. Roadmaps look different. Assumptions get challenged earlier. And innovation arrives not just faster, but stronger, because it has been pressure-tested by the very people who know how to protect it.

2.4.4 From Silos to Synchrony

Too often, MedTech feels like an orchestra warming up, everyone skilled, but not quite in tune. Engineering drives ahead. Regulatory checks after the fact. Quality steps in when something cracks. The music is there, but it is fragmented.

The SHIFT mindset asks us to aim for synchrony, not just coordination. Each function keeps its depth of expertise, but they work from the same score, with shared cadence. The result is not noise control, but actual music: progress that feels coherent, timely, and trusted.

2.4.5 Psychological Safety as the Soil

None of these sticks without psychological safety. SHIFT is less about frameworks and more about what happens when someone spots a problem. Do they speak up or stay quiet?

In too many rooms, people soften their words for political safety. They "note" concerns instead of challenging them. Over time, the system looks sturdy but becomes brittle. Real safety means people can escalate without fear of blame. Leaders model this by inviting challenge, rewarding candour, and showing that raising a risk is an act of care, not defiance.

When that happens, quality stops being a checkbox. It becomes a shared ethic, the way a team protects one another and, ultimately, the patient.

2.4.6 A Mindset for the Moment

SHIFT isn't a magic fix. It is a compass. It guides how we build, how we lead, and how we learn in a time when complexity is only accelerating.

Most importantly, it signals a transition: from defensive to proactive, from parallel functions to shared responsibility, from "checking the box" to "owning the outcome."

The chapters ahead will explore how to make this mindset structural, not just as a philosophy, but as an operating reality. But it starts here, with how we see one another, how we speak to one another, and how we choose to show up. Because culture isn't written in a policy. It is lived in the room.

2.5 From Barriers to Bridges - Enabling Convergence

2.5.1 The Illusion of Boundaries

Walk into almost any MedTech company and you will find extraordinary talent. Engineers stretch design possibilities.

Regulatory experts wrestle with global complexity. Quality leaders put guardrails in place to keep patients safe. But if you look past the activity, you will also see something harder to spot: invisible walls.

These walls aren't born of malice. They come from legacy roles, entrenched habits, and incentives that don't quite line up. The result is quiet fragmentation, teams moving in parallel, sometimes duplicating work, sometimes working at cross-purposes, and occasionally undermining each other without meaning to.

We often tell ourselves these boundaries protect focus and efficiency. In reality, they no longer do. In a world where MedTech grows more complex by the day, those walls block the very synchronisation we need and slow down the progress we can't afford to delay.

2.5.2 Convergence as a Strategic Imperative

Functional convergence is not an aspirational ideal, it is a strategic requirement. To deliver safe, innovative, and scalable medical technologies, innovation, quality, and regulatory affairs must move as one. Not in lockstep, but in harmony. This requires more than collaboration. It requires structural integration.

Convergence means early co-creation, not late-stage consultation. It means joint ownership of outcomes, not siloed accountability. It transforms the traditional product development model, from sequential handovers to continuous alignment. And it recognises that brilliance in one domain cannot compensate for fragmentation across domains.

The organisations that are realising this shift are not waiting for crises or audits to force convergence. They are choosing it: consciously, courageously, and systematically.

2.5.3 Integration by Design, Not Exception

In most organisations, cross-functional integration happens as an exception, usually when something goes wrong. But true convergence must be designed into the operating model. It begins with how teams are structured, how governance is enacted, and how success is measured.

Forward-looking companies now embed quality and regulatory functions in innovation teams from day one. These professionals are no longer seen as external advisors but as internal collaborators. Their input is sought during ideation, their insight shapes risk modelling, and their questions improve product resilience before a single design freeze.

When convergence is built into the system, the benefits compound. Risk is surfaced early, trade-offs are navigated transparently, and compliance becomes an embedded capability, not a last-minute scramble. This is not about adding steps. It is about replacing rework with readiness.

2.5.4 The Practical Enablers of Convergence

To build convergence, several practical enablers must be activated.

First, redefine roles. Move from functional depth alone to include translational fluency. Quality leaders should understand user empathy. Engineers should understand regulatory strategy. Cross-pollination of experience deepens insight and builds mutual respect.

Second, invest in digital infrastructure. Collaboration tools, real-time traceability systems, and unified documentation platforms are essential. These tools reduce redundancy, ensure visibility, and allow dispersed teams to operate as if they are in the same room, regardless of location.

Third, revise incentives. Too often, KPIs are function-specific and misaligned. The regulatory team is rewarded for low audit findings, the innovation team for speed, and the quality team for compliance. But these metrics drive division. Shared metrics, such as lifecycle value, audit-readiness at launch, or time to safe scale create shared focus and joint accountability.

Fourth, rethink governance. Governance is often associated with control, but in converged organisations, it becomes a mechanism for coordination. Stage gates evolve into synchronisation points. Reviews become forward-looking design sessions, not backward-looking audits. And decision rights are clarified not by hierarchy, but by contribution to enterprise value.

2.5.5 The Human Side of Integration

While systems and structures matter, convergence is also deeply human. It requires curiosity, empathy, and the willingness to operate beyond one's lane. It requires that leaders model humility, that teams share wins, and that feedback is welcomed as a form of respect, not critique.

Psychological safety again plays a foundational role. Cross-functional collaboration only thrives when people feel safe to disagree, ask naive questions, and raise early warnings. When people are rewarded for surfacing risk, not penalised for it. And when challenge is seen as a commitment to excellence, not a threat to authority.

It also requires time. Building trust across domains is not instant. But with sustained intention, cross-functional relationships mature into partnerships, and partnerships mature into shared identity. Over time, the question shifts from "who owns this?" to "how do we succeed together?"

2.5.6 From Parallel Paths to Shared Purpose

True convergence doesn't mean every function dissolves into the other. It means they move with shared cadence, shared purpose, and shared understanding. Each function retains its expertise, but that expertise is now in service of a collective outcome: safe, effective, and trusted medical technologies.

This shift, from parallel paths to shared purpose, is a profound one. It changes how decisions are made, how problems are solved, and how organisations respond under pressure.

When convergence is realised, meetings feel different. Conversations are faster, deeper, and more solutions-oriented. Escalations reduce. Surprises diminish. And most importantly, the patient, the person at the end of every decision, moves from abstraction to anchor.

This is not just an improvement in process. It is a restoration of purpose. Because when our functions converge, our intentions do too. We stop guarding our domains and start guarding each other's blind spots. We stop protecting turf and start protecting lives.

And that, ultimately, is the bridge that matters most.

2.6 Synchronisation as Strategy - Unlocking Advantage

2.6.1 Beyond Alignment: Entering the Era of Synchronisation

For years, organisations have spoken about "alignment" as the benchmark: making sure teams are on the same page, plans are roughly coordinated, and no one is pulling in opposite directions. But in today's MedTech environment, alignment alone is no longer enough. The speed, scrutiny, and complexity of delivering safe, compliant, and innovative technologies requires something far more fluid and responsive: synchronisation.

Synchronisation doesn't mean everybody agrees on everything. It means the enterprise moves with coherence with diverse teams in quality, regulatory, innovation, clinical, and commercial working to the same rhythm, anticipating each other's needs, and adjusting together. It is less about boxes on an organisational chart and more about flow. And in companies that achieve it, synchronisation becomes a competitive advantage competitors cannot easily imitate.

2.6.2 When Friction Fades, Velocity Emerges

Every leader has seen the drag that comes from misalignment. A simple task can balloon into weeks of friction. Feedback loops stall. Approvals stack up. Reviews turn into rework. Small missteps snowball into delays that no one anticipated.

Synchronisation cuts through that drag. Instead of sequential handovers, teams work side by side. Quality isn't the after-the-fact inspector, it is shaping robustness in real time. Regulatory isn't the final gate, it is a design partner. Innovation isn't just invention, it becomes about delivery, scale, and resilience.

The difference shows up in velocity, not because people are working harder, but because the resistance is gone. Products move from concept to submission to launch with fewer surprises. Issues surface upstream, where they are manageable. Lessons flow downstream, where they prevent repeats. And trust, both inside and outside the organisation, steadily compounds.

2.6.3 Predictability in an Uncertain World

The MedTech landscape is defined by uncertainty: shifting regulations, new customer expectations, evolving clinical realities. In that kind of environment, predictability is power. Synchronised organisations adapt faster, absorb shocks more gracefully, and maintain momentum even under pressure.

Consider the launch threatened by a last-minute regulatory interpretation. In a fragmented company, the reaction is chaos: urgent meetings, delay notices, and frustration all around. In a synchronised organisation, the change is absorbed calmly. Regulatory, quality, and engineering teams already share context. The impact is assessed quickly. Adjustments are made without panic. The launch stays on track.

That kind of resilience doesn't happen by chance. It is the by-product of embedded trust, ongoing cross-functional dialogue, and a systems-thinking mindset. In effect, synchronisation becomes the organisation's insurance policy, protecting timelines, reputation, and patient safety.

2.6.4 How Synchronisation Builds Market Trust

Regulators can tell when a company is aligned internally. Submissions are consistent. Data supports claims without contradiction. The narrative is coherent. That kind of credibility builds trust, and the regulatory journey becomes smoother, not because of shortcuts, but because confidence is earned.

The same dynamic plays out in the market. Hospitals and clinicians are quicker to adopt products from companies that deliver reliability again and again. Synchronisation ensures that field feedback isn't lost but acted upon and fed back into development. Updates are intentional, validated, and communicated with clarity. That reliability leads to fewer recalls, faster adoption, and lasting customer confidence.

Over time, that trust hardens into brand equity. It shapes loyalty. And in a crowded MedTech landscape, it becomes one of the sharpest differentiators.

2.6.5 The Human Advantage of Cultural Synchrony

Synchronisation isn't only about processes and governance. It is cultural. When teams move in rhythm, morale rises. Respect grows. The walls between functions soften. People feel less like isolated specialists and more like contributors to a shared mission.

In this kind of culture, the quality manager is no longer a lonely voice of caution. The engineer feels empowered instead of policed. Regulatory professionals are seen as partners in building trust, not as obstacles.

That human synchrony matters as much as the technical processes. It keeps talent engaged, fuels innovation, and helps teams ride out crises with composure instead of chaos. Culture becomes the glue that holds strategy, structure, and execution together.

2.6.6 Synchronisation: The Differentiator That Lasts

The next era of MedTech will not be won by invention alone. A clever algorithm, a novel implant, or a new delivery system will capture attention, but attention is fleeting. What lasts, what earns adoption, trust, and market share is the ability to deliver those innovations with speed, safety, and reliability, not once but repeatedly. That is the power of synchronisation.

Synchronisation builds companies that are fast without being reckless, precise without being paralysed. It creates organisations that are compliant and creative, rigorous and resilient, integrated and intelligent. In synchronised enterprises, every decision carries less friction, every iteration moves faster, and every launch lands with greater confidence.

And here is the true advantage: synchronisation cannot be copied overnight. Competitors can reverse-engineer your technology. They cannot easily replicate your culture, your trust, or the rhythm that lets your teams move as one. That is why synchronisation is not just how you break ahead in MedTech: it is how you stay there.

The future will favour companies that work in unison: organisationally, operationally, and ethically. Synchronisation does not just reduce risk, it amplifies resilience. It does not just improve process, it restores purpose. And in a field where the stakes are measured in human lives, synchronisation is more than a strategy. It is a commitment to every clinician, every regulator, and most of all, every patient we serve.

2.7 The Road Ahead – Courage, Clarity, and Cadence

2.7.1 Leading the Leap, Not Managing the Drift

Choosing synchronisation is not about convenience, it is about leadership. The systems most of us inherited are siloed and sequential, are designed for a slower, simpler era, and cannot keep pace with today's MedTech reality. Yet too often, organisations manage the drift instead of leading the leap. They patch over cracks with new checklists, double down on legacy processes, or wait until regulators or crises force their hand. But delay is no longer neutral. In today's environment, hesitation is a liability.

Leading the leap requires vision, but more importantly, it requires resolve. The leaders who succeed in this new era are those who recognise that the future will not wait, and who act with intention to meet it head-on. They do not wait for a crisis to prove the cost of misalignment. They create momentum through foresight, not fear.

2.7.2 From Intention to Integration

Transformation does not begin with a grand restructuring plan. It begins with small but deliberate integrations, starting now. A product team where innovation, quality, and regulatory work as joint architects rather than separate reviewers. Meeting rhythms where every voice is present early, not late. KPIs that measure lifecycle success instead of functional silos.

The first step is not scale. It is synchrony. Bring the right people together early and often. Make shared risk ownership normal, not exceptional. Replace functional scorecards with outcome-based measures that everyone recognises as meaningful: patient safety, time to market, post-market stability, and audit readiness.

That is how new habits take root. Not in sweeping declarations, but in the steady cadence of shared decision-making, cross-functional learning, and continuous reflection.

2.7.3 Finding Clarity in Complexity

In complex systems, clarity is more powerful than control. Leaders often assume that adding more process will increase predictability. But the opposite is usually true. Excessive control slows work, blurs accountability, and creates rigidity. Clarity, on the other hand, frees people to act with confidence.

Clarity means every team understands not only what to do, but why it matters. It means seeing how local actions ripple through the system. It means sharing a common understanding of risk, not just regulatory definitions, but the operational, reputational, and patient dimensions.

In synchronised organisations, clarity displaces ambiguity. Teams move decisively because they trust the alignment of purpose and direction. That kind of clarity is contagious. It shapes culture, simplifies strategy, and unlocks scale.

2.7.4 The Rhythm of Resilience

Change is not a single act. It is a rhythm. Transformation requires cadence: a steady, repeatable beat that signals we are moving together. This rhythm shows up in rituals, cross-functional design reviews, integrated post-market debriefs, and shared learning forums. Cadence turns aspiration into execution. It is what keeps momentum alive.

Resilient organisations don't sprint once and rest. They keep moving, adjusting as they go. When disruption comes, they respond with poise, not panic. They have built muscle memory across teams, allowing them to pivot without losing synchrony. That is the quiet power of cadence: it transforms synchronisation from a lofty idea into a lived discipline.

Without cadence, even talented organisations stall. Teams fall back into firefighting, projects lurch from crisis to crisis, and leaders mistake motion for progress. With cadence, however, trust compounds. Small wins stack. People begin to anticipate each other's moves the way musicians in an orchestra do, confident not only in their own part, but in the collective rhythm.

Cadence creates resilience because it trains organisations to move under pressure without breaking stride. It is what allows a company to absorb shocks, adapt to new regulations, and still deliver with consistency. In the long run, cadence is not just about speed. It is about endurance, the capacity to keep pace with complexity while staying aligned to purpose.

2.7.5 The Courage to Lead Differently

At the centre of all this lies courage. The courage to challenge traditions that no longer serve. The courage to elevate quality from a compliance checkpoint to a driver of contribution. The courage to release familiar hierarchies in favour of shared ownership. And above all, the courage to lead with humanity, to create spaces where people feel safe to challenge, to speak up, and to care.

Without courage, organisations retreat into comfort zones. Leaders hide behind process instead of purpose. Teams default to silence rather than risk speaking truth. Hierarchies harden, and innovation suffocates under the weight of caution.

With courage, the opposite happens. A regulatory leader invites engineers into strategy conversations, not just submissions. A quality manager raises a difficult concern and is thanked for foresight, not punished for delay. A senior executive openly admits what they don't know and asks for help, modelling trust rather than control. These moments may feel small, but they are the seeds of systemic change.

This is not abstract. It is deeply practical. Synchronised enterprises begin with synchronised leadership: leaders willing to admit what isn't working, humble enough to listen, and brave enough to start again. And courage, once visible at the top, cascades. It emboldens teams to take accountability, to surface risks early, and to pursue progress without fear.

Because in the end, courage is not just about standing against resistance. It is about standing for resilience, for trust, and for the people whose lives depend on the systems we build.

2.7.6 From Shift to System

The SHIFT mindset introduced here is not a slogan. It is a foundation, a move toward coherence in a world that too often rewards fragmentation. It is about embedding integration, not just talking about it. Turning feedback into action, not just collecting it. Building trust, not just expecting it.

Before moving to the iQSM model, pause for reflection. What is the current state of your environment? What role do you play in shaping change? Where can you close the gap between progress and process?

This is not just a call to reflect. It is a call to realign. Ask yourself and your teams:

- Are innovation, quality, and regulatory working from shared goals or parallel assumptions?
- Where do cultural undercurrents or legacy mindsets resist convergence?
- How are you bridging the gap between bold ambition and scalable, compliant execution?
- Which SHIFT pillar — Systems Thinking, Harmonisation, Integration, Feedback, or Trust — needs urgent attention in your organisation?
- Are you treating synchronisation as a strategy or as a patch when things break?

Now we move from mindset to model. In the next chapter, the Integrated Quality Synchronisation Model (iQSM) shows how these principles can be hardwired into the fabric of MedTech organisations.

Because the future will not belong to those who simply hope for harmony. It will belong to those who build it: intentionally, bravely, and together.

BUILDING THE SYNCHRONISED CORE - SYSTEMS, STRUCTURES & STRATEGY

The vision is clear. We have felt the urgency. We have seen what misalignment costs in recalls, delayed launches, and fractured trust. We have also glimpsed what is possible when synchronisation works, when quality, regulatory, and innovation move together, and progress flows without friction. The SHIFT mindset gave us the language. Now we need the machinery.

In MedTech, theories don't move products. What moves are submissions, risk files, protocols, training records, dashboards and ultimately, the devices placed in patients' hands. The environments are unforgiving: high-speed, high-stakes, and highly regulated. Vision without structure collapses into noise. Strategy without systems drains into burnout. That is why the next section matters. It is the backbone of this book, the scaffolding that makes synchronisation more than an inspiring idea. This section makes synchronisation how we work, how we decide, and how we endure.

Chapter 3: Blueprint for a Living System unveils the Integrated Quality Synchronisation Model (iQSM). This isn't a minor tweak to old QMS thinking. It is a new organising principle, a living framework that ties innovation, quality, and regulatory together from the start. Born from the collapse of silos, iQSM doesn't just manage compliance, it gives teams a compass for working in rhythm across functions, geographies, and product lifecycles.

Chapter 4: Global Harmonisation takes us into the regulatory theatre. MedTech is global by nature, yet too often local rules fracture global progress. Harmonisation isn't about watering things down, it is about coherence, about ensuring that success in one market doesn't sabotage speed or safety in another. We will explore how regulatory intelligence, foresight, and digital tools can turn cross-border complexity into competitive advantage.

Chapter 5: Living Systems reimagines the QMS itself. Too many companies treat their QMS like a static archive, a place audits visit, not a place teams live in. This chapter shows how to shift from passive paperwork to an adaptive system that breathes with the business. One where compliance is embedded in daily rhythm, training is a lever not a burden, and documentation is alive, not shelfware.

Chapter 6: The Innovation Execution Engine brings it down to the floor. Synchronisation only matters if it changes daily work. This chapter covers how frameworks become dashboards, how PowerPoints become team rituals, and how alignment translates into faster launches with fewer surprises. It is the anatomy of execution when vigilance and velocity share the same track.

Chapter 7: Governance Reimagined addresses the overlooked machinery: decision-making. Governance in many firms is still treated as control and sign-off. In synchronised organisations, governance is coordination. It provides clarity of roles, digital transparency, and accountability that accelerates rather than blocks. Done well, it is the nervous system that keeps the organism agile and coherent.

Taken together, these chapters form your operating system. Without it, synchronisation remains a hopeful slogan. With it, it becomes a competitive advantage, a cultural differentiator, and the engine of safer, faster, smarter innovation.

This is the moment where SHIFT stops being philosophy and becomes practice. The choice is stark: stay trapped in silos and watch progress fracture or build the systems that let teams move with courage, clarity, and cadence. The path ahead is not easy, but it is necessary. And it begins here.

CHAPTER 3

BLUEPRINT FOR A LIVING SYSTEM: INTRODUCING THE IQSM FOR MEDTECH EXCELLENCE

F or decades, excellence in MedTech has been pursued through control. Binders of procedures, elaborate validation matrices, and rigid phase gates, all designed to keep risk in check. These frameworks promised safety through structure, and in a slower, hardware-dominated world, they largely worked. But today, in an environment of rapid iteration, distributed global teams, digital health ecosystems, and shifting regulatory frontiers, their limits are clear. Control alone is no longer sufficient. Misapplied, it can become the very force that slows the agility and synchronisation needed to deliver safe, trusted innovation at speed.

MedTech does not need another static model. It needs a system that lives and breathes. One that grows with the organisation, senses shifts in its environment, and anticipates challenges instead of reacting after the fact. A system where quality, regulatory, and innovation teams do not run as parallel tracks but move like interdependent rhythms in the same composition.

This is where the Integrated Quality Synchronisation Model (iQSM) comes in, not as a cosmetic rebranding of the QMS, but as a fundamentally new architecture. iQSM shifts the organising principle from compliance to synchronisation, from static control to dynamic coordination. It is not about ticking boxes faster. It is about building a system that sees the whole, acts with intent, and adapts intelligently as reality shifts.

In this chapter, iQSM is introduced not as a theory on paper, but as a practical, implementable blueprint. It redefines how MedTech companies can structure excellence across the entire lifecycle, from concept to market release and through post-market learning. We will explore how it is built, what makes it distinct, and why its layered design offers both operational clarity and strategic resilience. We will look at its architecture, its guiding principles, and the mindset shift it requires.

The truth is simple: the world no longer rewards isolated perfection. It rewards those who can adapt without losing integrity. Move fast without losing safety. Innovate without compromising trust. iQSM is not a tool for specialists. It is the foundation for MedTech's next chapter.

3.1 Breaking the Silos: Why Fragmentation Fails In MedTech

3.1.1 A Legacy of Separation

To understand why integration matters, we first need to see how fragmentation became the norm. For decades, MedTech grew under a culture of separation: design focused on creativity, regulatory on compliance, and quality on oversight. These disciplines drifted apart not through malice but through habit, hierarchy, and history.

In the days of long hardware cycles and relatively simple regulatory expectations, this worked. Silos provided focus and gave each function clear lanes of responsibility. But today, devices are interconnected, software-driven, and governed by global rules that change faster than product roadmaps. Yesterday's structure cannot keep pace with today's speed.

What once gave clarity now creates clutter. What once defined roles now creates disconnection. And the costs are greater than most leaders realise.

3.1.2 The High Cost of Disconnection

Silos are not neutral. They create drag, duplication, and missed signals. Too often, the root cause of recalls or audit findings is not bad design, but weak integration. Labelling errors. Software validation gaps. Usability oversights. All symptoms of functions working in parallel without truly seeing each other.

I recall one case where an engineering team pushed a software update that cleared internal checks but skipped regulatory review. Quality was never looped in until complaints surfaced in the field. Nobody acted recklessly, but the absence of synchronised oversight turned functional excellence into systemic failure.

The true cost of silos is not only market delays or warning letters. It is the erosion of trust. It is teams losing momentum. It is the widening gap between what we intend and what actually reaches patients. Silos don't just slow us down. They shrink what we believe is possible.

3.1.3 Why Silos Fail the Future

MedTech today is defined by agility, complexity, and real-time feedback. The old logic, where quality followed design and compliance was layered on after innovation, no longer fits.

Success now depends on synchrony: systems that talk to each other, teams that work across functions, and information that flows without friction. The architecture of the future is not linear handoffs but living networks.

After a failed pre-launch review, I once told a senior executive:

> *"We didn't fail because we lacked talent. We failed because our talent couldn't see each other."*

The silence in that room said it all. This was not a failure of effort. It was a failure of connection.

3.1.4 A System in Transition

The shift from fragmentation to synchronisation is not only technical, but also cultural. It requires dismantling reward systems that celebrate functional heroics. It means measuring not just outputs, but collaboration and shared accountability. Above all, it requires designing for shared intent.

Synchronisation begins with a simple recognition: no single function can guarantee safety on its own. Innovation that is not regulatory-aware will stumble. Regulatory that is not innovation-savvy will stall. Quality that is not embedded in both will miss the mark.

As I often tell teams:

> *"When we dismantle silos, we do not lose control; we rediscover connection. And in that connection lies our greatest strength."*

3.1.5 The Path Forward

The end of silos is not the end of structure. It is the beginning of systems. Not chaos, but coherence. Not dilution of expertise, but alignment of purpose.

Synchronisation is no longer optional. It is the prerequisite for scalable, safe, and globally trusted innovation. The iQSM begins from this truth: excellence is no longer built in pieces. It is built in rhythm. And that rhythm starts here.

But rhythm requires leadership to set the tempo. It asks organisations to tune their instruments – innovation, quality, and regulation – so they no longer play over one another, but with one another. It challenges us to measure success not by the speed of a single team, but by the harmony of the whole.

This is not an abstract ambition. The companies that embrace synchronisation will ship products faster, recover from disruption quicker, and earn trust deeper. They will waste less energy fighting internal friction and spend more energy on patient impact. They will stop reacting to compliance crises and start anticipating opportunities.

The path forward is not easy, but it is urgent. Every delay in alignment is a risk to patients, a cost to teams, and a drag on innovation. Every step toward convergence is an investment in resilience, reputation, and real-world results.

And so, as we turn to iQSM, we are not adding one more framework to the shelf. We are laying down the operating system of the future, one designed not just to withstand complexity, but to thrive within it.

3.2 From Structure to System: Synchronisation as the Operating Model

3.2.1 Why Quality Systems Must Learn to Breathe

Quality systems were once built like scaffolds: rigid, predictable, designed to enforce stability. Their strength was control, catching errors, maintaining compliance, and enforcing uniformity across products and processes. For decades, that was enough.

But today's MedTech environment is unrecognisable. Devices are software-driven, updated in real time, and regulated across multiple fast-moving jurisdictions. Data streams in continuously from post-market surveillance, field performance, and patient feedback. In this reality, rigidity becomes fragility. Control alone slows more than it safeguards.

What organisations now require is a system that is not only engineered but alive, one that senses change, adapts quickly, and integrates signals across development, regulation, and quality. A system that doesn't just record events but anticipates them. Synchronisation is the principle that makes this possible: not a KPI, not a workflow tweak, but the organising model that lets quality, regulatory, and innovation departments breathe as one.

3.2.2 The Signals That Keep Us Alive

Think of synchronisation as the nervous system of a MedTech company. Just as the brain, spine, and heart operate in continuous dialogue, so too must our critical functions:

- Regulatory acts as the sensory input, scanning the horizon for risks, detecting changes in global requirements, and feeding the organisation with foresight.
- Quality provides the rhythm, setting standards, reinforcing discipline, and anchoring the consistency that gives confidence to patients, regulators, and clinicians alike.
- Innovation supplies the generative force, the creative muscle that pushes boundaries, designs solutions, and turns possibility into products.

Individually, these parts can function. But disconnected, they misfire. Regulatory senses danger too late. Quality beats out of rhythm with the rest of the body. Innovation surges forward but trips over unseen risks. The result is the organisational equivalent of paralysis: delays, recalls, and fractured trust.

Together, in synchrony, they form something greater: a living intelligence that perceives, decides, and adapts with purpose. Reflexes sharpen. Awareness deepens. Energy flows where it is needed most. Just as the body relies on its nervous system for survival, MedTech organisations rely on synchronisation to navigate complexity and deliver safely at speed.

This is the role of iQSM. Not a bolt-on framework. Not another compliance overlay. It is the connective tissue that turns fragmented functions into a coordinated whole, a system where signals are shared, decisions are clear, and the organisation moves as one.

3.2.3 Systems Thinking in Practice

The future belongs to companies that think in systems, not silos. That means moving beyond vertical towers of compliance or isolated corridors of creativity into closed-loop ecosystems where functions share accountability.

Systems thinking changes the questions leaders ask:

- Not just "What failed?" but "Why did this failure make sense in the system we built?"
- Not just "How do we comply?" but "How do we align compliance with innovation and patient trust?"

iQSM is the practical expression of this thinking. It replaces linear, check-the-box models with adaptive architecture. It braids quality, regulatory, and innovation into a single rhythm, each discipline distinct, but inseparable, each contributing to the same performance.

"A system becomes intelligent not when it reacts, but when it senses the need to act before disruption arrives."

3.2.4 Why Rigidity is Fragility

The past decade has made one lesson clear: rigidity is not resilience. The organisations that thrive are not those that hold still but those that move gracefully with change. They treat systems not as fortresses to defend, but as living frameworks that learn and adjust.

Synchronisation is a declaration of this new reality. It means designing organisations where regulatory oversight flows instead of obstructs, where compliance is embedded in daily work, and where quality is experienced as contribution, not constraint.

Letting go of legacy assumptions is not a loss, it is a liberation. iQSM shifts the centre of gravity from control and isolation to synchrony. In this model, each function listens, responds, and adapts in concert.

Through iQSM, we don't just build systems that function. We build systems that earn trust, absorb shocks, and enable innovation at the speed patients deserve. And through them, we do more than adapt to the future: we shape it.

3.3 The iQSM Unveiled: Blueprint for a Living MedTech System

3.3.1 A New Blueprint for a New Era

Traditional Quality Management Systems were built for a different world. They thrived in an era of predictable hardware, long development cycles, and relatively stable regulatory expectations. Their job was control: catch errors, enforce consistency, and keep organisations inside the lines. For that time, they worked.

But the world has changed. Devices are intelligent, connected, and continuously updated. Markets shift overnight. Regulations evolve faster than most teams can process. Risk no longer sits neatly inside a product's hardware; it is dynamic, distributed, and interdependent.

In this environment, tighter checklists are not the answer. What is needed is a new operating system. The iQSM provides that foundation. It reframes quality as an active force, not a reactive function. It positions regulatory as a source of foresight, not just documentation. And it embeds innovation into the very core of governance, not as an afterthought.

iQSM is not a bolt-on framework. It is the architecture that threads together three critical capabilities of modern MedTech: regulatory precision, quality intelligence, and innovation velocity. It replaces rigidity with adaptability. It turns structure into flow. And most importantly, it replaces functional isolation with synchronisation.

3.3.2 The Living Spine: Three Interdependent Layers

At the centre of iQSM is a three-layer structure, a living spine of synchronised intelligence. Like a body's backbone, it provides both stability and movement. Each layer plays a distinct role, but only together do they allow the organisation to sense, adapt, and grow. Remove one, and the body falters. Strengthen all, and the system moves with both speed and confidence.

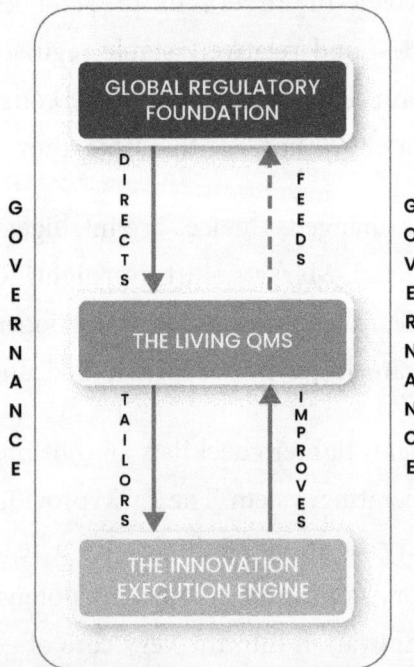

Integrated Quality Synchronisation Model

Layer 1: Global Regulatory Foundation: This is the external interface of the organisation, its eyes and ears on the world. It continuously absorbs regulatory updates, new standards, and shifting market requirements across geographies. Its modular design spans domains such as clinical evidence, labelling and IFU, cybersecurity, software, human factors, submissions, and post-market surveillance.

What makes this foundation powerful is its modularity. Updates can be absorbed without destabilising the whole, much like a spine can flex without breaking. It becomes the single source of regulatory truth, harmonised across geographies and translated into a language the enterprise can act on. For teams buried in daily deadlines, this foundation is more than compliance, it is foresight. It prevents surprises from blindsiding projects and ensures decisions are informed by the best intelligence available.

Layer 2: The Living QMS: If Layer 1 is the organisation's senses, the Living QMS is its nervous system. This is where regulation becomes action. It translates modular requirements into tailored, risk-based processes, drawing from ISO 13485, ISO 14971, IEC 62304, Agile, Lean, and Six Sigma. Its strength lies in adaptability.

Traditional QMS frameworks often become bloated archives where process is confused with paperwork. The Living QMS rejects this. It tailors process maps to product class, risk profile, and market strategy. These tailoring rules serve as practical blueprints, giving teams pre-approved pathways they can follow with confidence, avoiding overburden where it isn't needed. Most importantly, it is never static. It listens, learns, and evolves from performance data, team insights, and post-market feedback. In doing so, it shifts compliance from a drag on innovation to a driver of synchrony.

Layer 3: The Innovation Execution Engine: This is where strategy becomes movement. Here, innovation teams are not asked to decode global regulatory complexity, as it has already been absorbed upstream. They operate inside processes already embedded with regulatory and quality safeguards. Risk management, documentation, and compliance are integrated into their workflows, not bolted on at the end.

This frees teams to focus on creativity, iteration, and delivery, knowing that the system itself protects safety and compliance. In practice, this means fewer recalls, faster launches, and products that scale without unravelling. Innovation becomes not only faster, but safer. Not only leaner, but more trusted. The engine gives organisations the confidence to push boundaries without fear of losing control.

Together, these three layers form the living spine of iQSM. Each layer strengthens the others, creating resilience that no single function could achieve alone. But the model is only complete when wrapped by a fourth element: **Governance**.

Governance is the intelligent shell, not rigid oversight, but orchestration. It is the connective tissue that ensures signals move between layers, that accountability is distributed, and that insight is translated into action. Done well, governance sustains synchronisation not by enforcing control, but by cultivating clarity. It acts less like a gatekeeper and more like a conductor, ensuring the entire system moves in rhythm.

The result is not just compliance, but coherence. Not just systems, but symphonies.

3.3.3 A System Designed to Flow Both Ways

What makes iQSM distinctive is not only its structure, but its directionality. Most legacy systems are linear: requirements flow down, documentation flows sideways, and feedback, if it moves at all, trickles up too late to matter. iQSM is different. It is built for continuous flow. Not one-way. Not occasional. But a living loop.

Regulatory updates feed into the Living QMS. The Living QMS translates those requirements into execution pathways for product teams. Innovation, in turn, generates signals that flow back, updating risk files, refining regulatory strategies, and improving processes. This bidirectional loop is not an add-on. It is the design.

Consider cybersecurity. When new expectations emerge, the Global Regulatory Foundation absorbs the change immediately. The Living QMS converts it into updated workflows, training, and controls. These cascade into execution, so teams keep moving with clarity instead of confusion. What could have been a disruptive surprise becomes a smooth adjustment.

Now imagine the reverse. A field team flags a usability issue in a diabetes device. Instead of disappearing into a complaint database or dying in a departmental queue, that signal loops back. SOPs evolve. Design inputs are sharpened. Regulatory strategies are recalibrated. The loop doesn't just close, it strengthens. The system doesn't just obey, it learns.

And this is the crucial truth: shortcuts around compliance never create this kind of coherence. Strip out quality for speed and fragility creeps in. Delays resurface as rework. Trust erodes with recalls. What looked efficient collapses into crisis.

The three layers of iQSM are not stacked in hierarchy, but connected in harmony, orchestrated to deliver both velocity and vigilance. Because living systems don't move in straight lines; they pulse, loop, and adapt. iQSM mirrors that natural intelligence. It is not a framework to be endured, but a rhythm of coherence, ensuring progress is not only fast, but safe. Not only compliant, but confident.

3.3.4 The Blueprint Becomes Reality

iQSM is not aspirational. It is an implementable architecture for real-world complexity. It provides structure without rigidity, governance without obstruction, and compliance without delay. It enables quality and regulatory to act as embedded partners, not late-stage checkpoints. It allows teams to deliver with speed and clarity, not confusion and risk.

> *"Excellence is no longer a retrospective label; it is built in from the start."*

The critical question for MedTech organisations is no longer "Are we compliant?" but "Are we coherent?" Not "Can our systems control?" but "Can they learn?" Not "Do we meet standards?" but "Do we move in rhythm with change?"

iQSM offers a path forward, one built on responsiveness rather than rigidity, on synchronisation rather than separation. Through it, organisations do not just protect their future, they shape it.

And in that shaping, they build more than systems. They build resilience. They build trust. And ultimately, they build a MedTech world capable not only of surviving complexity, but of leading within it.

3.4 The Five Pillars of iQSM: Principles for Built-In Excellence

3.4.1 The DNA of a Living System

A system doesn't live just because it moves. It lives because it has purpose, because it knows why it moves. It senses, adapts, and evolves, not through rigid commands, but through guiding principles.

In iQSM, these principles are not slogans for posters or vision decks. They are the operating DNA, the code that determines how decisions are made, how risks are managed, and how synchronisation is sustained when pressure rises.

Policies alone will not carry a company through the next decade of MedTech complexity. Principles will. They ensure synchronisation isn't a one-time workshop or a lucky moment of alignment, but a discipline of working together. They are what keep the system alive when markets shift, when regulators tighten expectations, and when products evolve faster than processes.

In iQSM, this DNA is expressed through five interdependent pillars:

- Modularity by Design
- Tailoring Without Compromise
- Feedback as a Lifeline
- Governance as an Enabler
- Compliance That Breathes

Together, these pillars give iQSM both its structure and its agility, the bones and the muscle, allowing it to stand firm while staying responsive.

3.4.2 Modularity by Design

In MedTech, "one-size-fits-all" is not just lazy, it is dangerous. iQSM begins with a modular regulatory foundation. Instead of one monolithic system, it breaks compliance into clearly defined modules: clinical evidence, labelling, cybersecurity, submissions, post-market surveillance, and more.

This modularity is decisive. It means updates can be made quickly in one area without destabilising the entire framework. It creates clear ownership and scalable compliance. Each module evolves at its own pace, keeping step with changing standards or authority expectations, while still feeding into a harmonised global foundation.

Think of it like a well-designed operating system: every module can be updated, but the whole device still works seamlessly. Modularity is what allows iQSM to breathe with the market and to stay coherent while flexing with complexity.

3.4.3 Tailoring Without Compromise

Every product carries its own fingerprint – its own risk class, clinical pathway, and target market. And yet, in many organisations, those unique fingerprints are pressed through the same rigid mould. A Class I device is forced to march through the same documentation labyrinth as a high-risk implant. A software update is slowed by protocols designed for physical manufacturing. The result? Innovation is suffocated, and teams are left frustrated, completing activities that add paperwork but no real value.

iQSM challenges this. Tailoring in this model is not corner-cutting; it is structured, deliberate, and principled. Product teams apply predefined tailoring rules based on classification, intended use, and geography. These rules make clear what is essential, what can be adapted, and what is not required. Crucially, they are agreed upfront by cross-functional leaders, so no one is paralysed by ambiguity or forced to "negotiate compliance" in the middle of development.

This is agility with guardrails. It removes waste without weakening rigour. It gives teams permission to move fast with confidence, not because they are ignoring the rules, but because the rules were designed for their reality. Tailoring transforms the system from a blunt instrument into a precision tool.

In a simple term:

> *"Instead of treating every product like it*
> *needs a fortress, tailoring ensures each one*
> *gets the protection it truly needs."*

3.4.4 Feedback as a Lifeline

Most systems report. Few systems truly listen. iQSM is built to listen, continuously.

In many organisations, feedback is treated like a backlog item: a CAPA report at year-end, or an audit finding that surfaces months after the issue began. By then, the signal has gone cold, and the cost of correction has multiplied.

In iQSM, feedback is not an afterthought. It flows constantly, from usability testing in the lab, from field complaints, from shifting regulatory expectations, from training outcomes, and from real-world product performance. Every piece of feedback is treated as a signal, a pulse in the system, moving upstream and downstream, shaping design inputs, updating training, refining risk files, and improving processes.

A living system not only learns. It remembers. Feedback in iQSM is not a post-mortem report to explain failure. It is a survival instinct, as essential as oxygen, the mechanism that keeps the organisation alert, adaptive, and resilient.

Because in MedTech, it is not the company with the most data that survives. It is the company that can hear the faintest signal and act before it becomes a shout.

3.4.5 Governance as an Enabler

In too many companies, governance feels like the slowest room in the building, the place where projects stall, signatures pile up, and decisions get lost in committee. By the time approval comes, momentum has already leaked out.

iQSM flips that script.

Here, governance isn't control for control's sake. It is coordination. It is the connective tissue that holds the system together without choking it. Done right, governance keeps accountability clear, ensures signals move across functions, and helps decisions rise with rhythm, not in a last-minute scramble.

Think of it less as a gatekeeper and more like a conductor in an orchestra. The conductor doesn't play every instrument, but they keep the tempo, cue the entries, and make sure the music doesn't fall apart.

Good governance is like good health. You hardly notice it when it is working, but the moment it breaks down, everything else struggles. In iQSM, governance isn't the stamp that appears at the end of the process. It runs alongside the work itself, creating flow instead of friction, clarity instead of confusion, and speed without panic.

3.4.6 Compliance That Breathes

This is the hardest shift and the most liberating. In most organisations, compliance feels like a hurdle: the audit scramble, the late-night document chase, the checklist that lands after the design is already done. Everyone survives it, but few see it as something that actually helps.

iQSM flips that mindset. Compliance here isn't a burden. It is a capability, something you can trust to carry you forward.

Living compliance starts with one assumption: change is constant. New standards, new risks, new markets – the system is built to expect them. It doesn't freeze in a "final state." It scans, adjusts, and folds updates into daily work without derailing momentum. Teams don't experience it as paperwork dropped from above, they experience it as the way work naturally flows.

We have all seen what happens when compliance is treated as an afterthought: speed turns fragile, trust evaporates, and launches stumble. But when compliance becomes instinct, the opposite happens. It no longer slows the organisation down, it gives people confidence to move faster, knowing the guardrails are already there.

Living compliance is not about fear of failure. It is about freedom to act with clarity. That is when speed and trust can finally sit side by side.

3.4.7 The Interdependence of Strength

These five pillars – modularity, tailoring, feedback, governance, and living compliance – aren't a menu. You don't pick one or two and hope the rest takes care of itself. They only work as a set. Remove one, and the whole system wobbles. Modularity without feedback turns rigid. Feedback without governance becomes noise. Tailoring without living compliance invites risk. It is the interplay that gives the system rhythm.

That rhythm is what makes iQSM more than another framework. It feels less like a checklist and more like a heartbeat, something that keeps the organisation alive, moving, and in sync.

And here is the real difference: in this model, oversight doesn't slow innovation, it sharpens it. Regulation isn't tolerated at the margins, rather it is embedded at the core. Quality doesn't stand at the gate, it runs in the bloodstream.

Excellence, in iQSM, isn't stitched together from fragments. It is built in rhythm, the rhythm of teams that trust each other, of systems that learn, and of organisations that know how to move fast without losing their balance.

3.5 Laying the Foundation: Building for Sustainable Synchronisation

3.5.1 The Call for a Different Kind of System

Many MedTech organisations today are standing at a threshold. On one side is the comfort of the old: linear, department-led systems where everything is routed through documents and signatures. On the other side is the reality of now: devices that are smart, connected, and constantly evolving, markets that demand speed, transparency, and adaptability. Between those two worlds, the ground is cracking. Delays pile up. Trust thins. Opportunities slip through the gaps.

The iQSM isn't a patch across that fault line. It is the bridge. And that bridge can't be crossed with tweaks and cosmetic fixes. It asks us to rethink what a MedTech operating system is for. Not a library for inspectors, but a framework that keeps the business alive. Not a tower of procedures, but a living foundation where integration comes first.

The foundation of that system isn't made of binders. It is built on intent:

- The intent to align before crises force it.
- The intent to embed quality at the source of creation.
- The intent to turn regulation into readiness, and compliance into confidence.

3.5.2 A Company at the Crossroads: Inside a Mid-Sized MedTech Company

Picture a mid-sized diagnostic company with a proud engineering culture but a fractured operating rhythm. Quality ran on inherited SOPs: heavy, slow, rarely questioned. Regulatory was looped in late, often when designs were already frozen. Innovation teams sprinted ahead, only to be pulled back months later for rework. Feedback came sporadically, usually triggered by a failure, and rarely travelled far.

The breaking point came during a European market submission for their flagship AI-driven monitoring device. Reviewers found traceability gaps, mismatched files, and fragmented evidence chains.

The product itself was strong, but the system holding it together wasn't. The near-failure didn't come from bad science or weak design. It came from misalignment.

The executives could have patched the gaps. Instead, they took a bigger leap. They dismantled the brittle QMS and rebuilt their model using iQSM principles. A modular Global Regulatory Foundation was set up as the single source of regulatory truth. A Living QMS translated those requirements into risk-based blueprints, tailored to product class and market. Innovation teams didn't have to decode complexity anymore, rather they followed a map already wired with compliance.

Within a year, the company launched its next-gen platform across four major markets, on time and in compliance. But the real win wasn't the launch. It was the cohesion. Engineers, quality leads, and regulatory staff began speaking the same language. Regulatory wasn't a late-stage gate, but a strategic voice in design. Quality stopped being the rule-enforcer and became a performance partner. And innovation didn't brace for interference, it thrived because alignment was baked in.

The system didn't just comply. It breathed.

Breathing systems aren't built from policies alone. They are built when connection travels faster than confusion, and when trust flows more freely than fear.

3.5.3 Why Leadership Makes or Breaks It

No system synchronises on its own. Even the best frameworks collapse without leadership to hold them. Building iQSM isn't a quality project. It is a leadership act.

That means questioning legacy metrics that reward siloed success. It means letting go of the comfort of separation and leaning into the messiness of connection. Leaders have to do more than fund the shift, they have to model it. Invite regulatory into design reviews. Make risk a shared language across functions. Judge decisions not by departmental wins but by system-level impact.

In practice, this is hard. It means sitting in meetings that feel slower at first, because all the voices are at the table. It means resisting the urge to retreat into silos when the pressure mounts. It means choosing coherence over convenience.

The leader's job isn't to control the system. It is to nourish it. To make sure that when complexity accelerates, the organisation leans into synchrony instead of snapping back into silos. The real test isn't whether you can enforce alignment once. It is whether you can sustain it under pressure.

3.5.4 From Foundation to Future

iQSM is not something you "install" and leave behind. It is not a toolkit or a binder on a shelf. It is a way of thinking and working that has to be lived daily, adjusted constantly, and proven in practice. It doesn't promise perfection. What it promises is readiness. Coherence. The assurance that excellence is not bolted on at the end but built in from the very first step.

This chapter has laid the foundation: the end of silos, the three-layer living spine, the five pillars that keep it alive, and the leadership required to sustain it. But synchronisation within a single organisation is only the beginning. The bigger challenge lies beyond the walls of any one company: synchronising across regulators, across borders, across markets.

That is where we turn next, from internal rhythm to global resonance. From building coherence inside one enterprise to creating it across the MedTech world.

The system is alive. The foundation is set. The question now is whether the world is ready to move with it.

Before you turn the page, ask yourself and your teams:

- Is your quality system a living framework that adapts and guides, or just a static rulebook you work around?
- Where do silos still shape your processes, and what would it take to break them without breaking trust?
- Do your functions share the same definition of "good," or are they still scoring different games?
- Is synchronisation part of your daily rhythm, or only something you only reach for in a crisis?

With iQSM unveiled, the next step is turning concept into practice. The chapters ahead take us deeper into each of its three core layers, not in isolation, but as interdependent parts of a living system:

- The Global Regulatory Foundation: absorbing and translating global change into a harmonised source of truth.
- The Living Quality System: turning regulation into rhythm, synchronising processes with adaptability and clarity.
- The Innovation Execution Engine: converting strategy into motion, giving product teams speed with safeguards.

Surrounding them is Governance, the enabling shell that sustains alignment, flow, and accountability without suffocating innovation. Together, these layers show not only how iQSM breathes, but how it can be tailored, implemented, and sustained in the real world.

We begin at the top with the Global Regulatory Foundation. In Chapter 4: Global Harmonisation, From Fragmentation to Regulatory Flow, we will see why regulatory convergence an aspiration is no longer but an operational imperative. We will explore the modular architecture that underpins this foundation, make the case for regulatory flow over fragmentation, and show how alignment across borders can unlock both compliance confidence and market agility.

The future of MedTech will not be built in isolation. It will be built in rhythm, in step with a world that is more interconnected, more complex, and more synchronised than ever before.

CHAPTER 4

GLOBAL HARMONISATION: FROM FRAGMENTATION TO REGULATORY FLOW

In today's MedTech landscape, every product is born global. A diagnostic kit developed in Europe may need approval in the U.S. within months. An AI-driven surgical tool designed in California will almost certainly be marketed in Asia. Even the smallest innovation rarely stays within the borders where it was conceived.

But the path these products must travel is rarely smooth. Each jurisdiction brings its own frameworks, philosophies, and expectations. What organisations face is not just a patchwork of rules, but a fragmentation of logic – requirements that overlap in some places, contradict in others, and rarely move at the same pace. The consequences are real: delayed access for patients, rising operational costs, and regulatory and quality teams stretched thin as they reconcile the same product across divergent standards.

Ask anyone who has tried to prepare simultaneous submissions for the EU MDR and U.S. FDA: it is not simply extra work, it is double thinking, double documentation, and double stress. The science doesn't change, but the story has to be told differently, sometimes in ways that confuse rather than clarify.

This is why harmonisation is no longer optional. It has become a condition of survival. Organisations that treat global regulation as a set of isolated hurdles will always be reactive, always behind. Those that move from fragmented compliance to strategic coherence, however, unlock speed, trust, and resilience.

Within iQSM, this imperative takes form in the Global Regulatory Foundation, the system's compass and stabiliser. It doesn't flatten nuance or erase local differences. Instead, it absorbs shifting requirements, interprets them, and structures them into modular intelligence that can be used across the enterprise. Rather than a static library of documents, it becomes a living map: one place where clinical evidence, cybersecurity expectations, labelling requirements, and post-market surveillance obligations can be accessed, updated, and acted upon.

The alignment of frameworks such as EU MDR and IVDR, FDA's QMSR, ISO 13485, and MDSAP is not about making every market identical. That is neither possible nor desirable. Harmonisation is about building bridges through structured mapping, intelligent adaptation, and operational fluency. It means developing a regulatory language that teams can speak across borders while still respecting regional dialects.

Most importantly, alignment enables innovation to move at the speed of need. A breakthrough therapy or diagnostic should not stall for years simply because each regulator asks the same questions in slightly different ways. Harmonisation transforms duplication into flow, risk into readiness, and effort into acceleration.

This chapter lays out not only the perspective but also the pathway. Grounded in real-world practice, powered by digital capability and modular intelligence, it shows how regulatory foresight can transform fragmentation into flow.

Because in the end, the future of MedTech will not be led by those who are merely compliant. It will be led by those who are synchronised: organisations that can move many systems with a single, trusted rhythm.

"Where alignment begins, acceleration follows."

4.1 The Case for Harmonisation

4.1.1 A World Divided Cannot Innovate Together

Picture a MedTech company on the verge of launching a breakthrough diagnostic platform. The science is solid, the validation is complete, and clinicians are eager to use it. On paper, it is ready to change lives on multiple continents.

But the road to market is anything but smooth. In Europe, it needs a CE mark under EU MDR. In the U.S., it must fit into FDA's 510(k) predicate logic. In Canada, localisation requirements add friction. In Asia-Pacific, expectations diverge again, from cybersecurity demands to human factors documentation.

The product is singular, yet the regulatory map it must navigate is fractured and inconsistent.

This isn't a theoretical burden. It is a strategic choke point. Fragmentation multiplies cost, complicates compliance, and slows patient access. Worse, it creates hidden risk: duplicated testing, mismatched documentation, disjointed evidence trails. In today's

MedTech environment, where speed, trust, and global reach define competitiveness, fragmentation is no longer tolerable. Harmonisation is not just efficiency. It is a business necessity. And in a field where patient safety is on the line, it is also a moral obligation.

4.1.2 From Compliance Chaos to Strategic Clarity

True harmonisation is not about creating uniformity for its own sake. It is about aligning regulatory intent across borders and enabling companies to act with coherence. Frameworks like EU MDR, FDA's QMSR, ISO 13485, and MDSAP are each built on legitimate and critical foundations. But without alignment, they pull organisations in divergent directions, creating costly duplication and unnecessary ambiguity.

Harmonisation invites clarity: regulatory, operational, and strategic. It enables global MedTech firms to design once, validate once, and deploy across markets without sacrificing rigour. Instead of rewriting the same clinical narrative for five audiences, teams can focus on innovation and evidence generation. Quality leaders can embed control upstream. Regulatory affairs can shift from firefighting to foresight. And most importantly, patients get access to life-saving technologies without systemic delay.

> *"Harmonisation is not a theoretical construct; it is operational sanity and strategic foresight in action."*

4.1.3 Technology and Trust: The New Pillars of Unity

Harmonisation is catalysed by two powerful forces: technology and trust. Digital platforms are no longer just tools for submission tracking, rather they are synchronisation engines. They house multi-jurisdictional requirements, map modular content to country-specific needs, and trigger intelligent alerts when regulatory landscapes shift. Machine-readable dossiers, UDI databases, and real-time dashboards now form the digital nervous system of globally harmonised compliance.

But without trust, even the best systems fail. Trust among regulators, built through transparency and collaboration. Trust within organisations, built through alignment and shared understanding. Trust with patients and healthcare providers, built through consistent performance and safety across borders.

In a world increasingly driven by complexity, trust and technology together create the connective tissue that holds harmonisation in place.

4.1.4 The Strategic Imperative

This is not a future ambition, rather, it is a current demand. As AI-driven diagnostics, software as a medical device (SaMD), and platform-based therapies blur traditional regulatory lines, global harmonisation becomes more than an advantage, it becomes survival. The pace of change will not slow. Those who invest in harmonised foundations today will be tomorrow's frontrunners.

Harmonisation is a strategy. It is an operating model. It is a cultural discipline. It shifts regulatory from the backend of development to the frontend of innovation. And in the context of the iQSM, it is institutionalised within the Global Regulatory Foundation, the topmost layer that filters complexity and anchors clarity.

Let us now explore the consequences of ignoring this imperative, the pitfalls that unfold when global fragmentation is left unchecked.

4.2 When Fragmentation Fails: Pitfalls of Global Launches

4.2.1 One Product, Many Barriers

In a world where patients wait across borders and technologies evolve in real time, the idea of launching a single product across multiple geographies should be a seamless process. Yet, for many MedTech firms, global expansion feels like navigating a regulatory obstacle course. A device that meets rigorous standards in one region often faces redundant or divergent demands in another. Differences in clinical evidence expectations, dossier structure, labelling protocols, and surveillance obligations create a web of friction that slows even the most prepared organisations.

This is not a matter of minor localisation; it is a structural inefficiency that disrupts business continuity and patient access. A single product can face a dozen different interpretations of what constitutes "safe and effective." The complexity compounds when confronted by changes in regulatory guidance, political shifts, or delayed recognition agreements. What should be a straight path becomes a maze.

4.2.2 A Case Delayed: The Cost of Misalignment

Consider a fictional but representative scenario: a MedTech company develops a next-generation wearable for cardiac monitoring. It receives early CE mark approval under EU MDR and begins market deployment. Emboldened by success, the company turns to the

US, Canada, and Australia. The regulatory dossiers are modelled on European documentation, assuming minor adjustments will suffice.

The assumptions unravel quickly. The US FDA challenges the predicate device and requires more robust software validation. Canada mandates bilingual labelling with specific localisation criteria. Australia raises concerns over cybersecurity protocols and post-market surveillance readiness. Each region demands changes that are logical in isolation but disconnected in sequence.

Approval delays stretch into quarters. Development teams are pulled from future pipelines to retro-engineer solutions. Investor confidence wavers. A competitor seizes the window and reaches key markets first. What began as a product poised for global impact becomes a cautionary tale of regulatory misalignment.

4.2.3 The Human and Organisational Impact

Behind every delay lies a team under strain. Engineers forced to rewrite documentation for different templates. Quality professionals stuck in endless loops of SOP modification. Regulatory affairs teams bridging gaps that systems were never designed to span. Time that should be spent advancing innovation is instead consumed by translation: of formats, of frameworks, and of expectations.

Worse, this burden rarely remains invisible. It creates fatigue, fractures cross-functional collaboration and dampens morale. Teams begin to view regulatory engagement as an unpredictable variable rather than a strategic asset. Blame replaces trust. Frustration replaces foresight.

And for patients, the cost is measured not in days, but in access lost. Every additional barrier adds time. And every missed month may mean another patient without access to a life-saving technology.

4.2.4 The Strategic Imperative for Cohesion

Global ambition must be matched by systemic preparedness. Fragmentation is not just a compliance inconvenience, it is a strategic threat. It undermines time-to-market, inflates cost projections, and erodes the confidence of both teams and stakeholders. While regional specificity will always exist, the path forward lies in anticipating divergence, not reacting to it.

This is precisely where the iQSM model becomes indispensable. With its Global Regulatory Foundation as the top layer, organisations are empowered to absorb complexity and translate it into coordinated execution. Instead of duplicating effort, they map alignment. Instead of localising in silos, they design with foresight. Harmonisation, embedded early, transforms fragmentation from a barrier into a manageable variable.

4.2.5 From Reactive to Proactive

The future of global MedTech leadership belongs to those who prepare, not those who scramble. This means regulatory inputs must be woven into the very fabric of product design, risk management, and clinical strategy. Global launch planning should begin not at submission, but at concept. Product documentation must be structured with modularity in mind. Training, evidence, and traceability systems must be designed to flex seamlessly once, effectively, and at a global scale.

Harmonisation is not about removing difference. It is about removing surprise. By institutionalising foresight, aligning early, and treating global scale as a design principle, not an afterthought, organisations move from chaos to confidence.

Synchronisation at the regulatory level is not a nice-to-have, it is the engine of global relevance. And the companies that lead will not be those with the most resources, but those with the most coherence. Let us now explore how to build that coherence intentionally, through frameworks that bridge the divide and make alignment real.

4.3 Bridging the Divide: Frameworks for Regulatory Alignment

4.3.1 When Systems Speak Different Languages

Regulatory systems across the globe are not in conflict. Rather, they express shared principles – patient safety, product efficacy, and risk-based oversight – through different dialects. Each framework carries unique emphases: where one region prioritises clinical validation, another scrutinises manufacturing controls or cybersecurity. The divergence is not philosophical, it is structural and semantic.

This divergence creates operational drag. A technical file approved in Europe under MDR cannot be submitted unchanged to the US FDA without navigating a labyrinth of formatting conventions, predicate justifications, and device classification nuances. Meanwhile, ANVISA, Health Canada, and the TGA introduce their own interpretative layers. The core science may be identical, but the translation is laborious and, too often, duplicative.

True harmonisation does not erase these differences. It creates fluency between them. The solution is not to force-fit compliance, but to establish operational frameworks that make equivalence actionable, traceable, and repeatable across geographies.

4.3.2 Mapping the Landscape: Recognising Structure and Spirit

Every regulatory framework has a structure: clauses, forms, checklists, and a spirit, the intention behind them. EU MDR is shaped by public trust and heightened post-market vigilance. FDA QMSR is anchored in design control and manufacturing discipline. ISO 13485 standardises global quality practices with flexibility in mind. MDSAP offers multi-region oversight with rigorous audit preparation. Understanding the spirit behind the structure is what enables strategic alignment.

Leading organisations build side-by-side regulatory matrices, mapping where requirements converge, where interpretation differs, and where local adaptations are necessary. These matrices go beyond compliance checklists, they function as strategic artefacts. They inform design inputs, dictate evidence generation, shape labelling architecture, and guide post-market readiness. Regulatory intelligence becomes an active map, not a retrospective summary.

The Global Regulatory Foundation layer of the iQSM sits at the heart of this capability. It enables modular documentation that can scale, adapt, and localise without redundancy. It provides the scaffolding for regulatory design that anticipates divergence and neutralises its disruptive potential.

4.3.3 Case in Flow: When Frameworks Enable Confidence

At a global diagnostics company expanding into five major markets, leadership made a pivotal choice: they embedded a harmonisation logic into product development from day one. Instead of allowing each regional regulatory lead to interpret requirements independently, a central regulatory alignment function was created. This team worked in concert with R&D, clinical, and quality counterparts to codify core documentation in a modular structure.

Design history files were structured around shared global requirements, with annexes tailored to jurisdiction-specific needs. Risk management documentation was crafted to meet both ISO 14971 and FDA expectations simultaneously. Clinical evidence planning integrated MDR's stringent performance evaluation criteria alongside predicate logic needed for a 510(k). When regional submissions commenced, each dossier drew from a single source of truth: refined but never reinvented.

The result was not only speed to submission, but clarity across functions. Review cycles were shortened. Audit readiness improved. And team fatigue, so common in global launches, was replaced with confidence, enabled by an aligned framework that removed chaos from complexity.

4.3.4 Harmonisation as Leadership Discipline

Framework alignment is not a regulatory task alone. It is a leadership discipline. It requires investment in systems, training in regulatory literacy across functions, and a mindset that treats alignment as a strategic lever. The most successful MedTech organisations no longer view harmonisation as an exercise in translation. They see it as an operating model, one that connects regulatory strategy to business outcomes.

In the iQSM model, this discipline is structural. The Global Regulatory Foundation is not a document repository, it is a living interface that unites global foresight, regulatory mapping, and change control into a single, synchronised capability. It ensures that teams speak a common language even when jurisdictions differ. It enables product, quality, and regulatory groups to design together, rather than adjust in isolation.

As global regulation continues to evolve, with AI, sustainability, and post-market digital surveillance reshaping the landscape, this capacity to bridge differences will define the pace and confidence with which innovation reaches the world. Alignment is no longer an option, it is the discipline of those prepared to lead.

Let us now turn our focus to regulatory intelligence and foresight, the strategic lens through which harmonisation becomes not only possible, but predictive.

4.4 Looking Ahead: Regulatory Intelligence and Foresight

4.4.1 Seeing Beyond Compliance

Regulatory excellence in the past was often a reactive discipline. It meant responding quickly to new guidance, updating documentation post-factum, and ensuring that products cleared the necessary hurdles just in time. That approach once sufficed in a world where changes were infrequent, products were largely hardware-based, and market strategies were sequential. Today, that world no longer exists.

In its place is a dynamic, non-linear reality. Regulatory expectations now evolve continuously, shaped by rapid innovation, geopolitical pressures, public health events, and digital convergence. Guidance no longer trickles in, it accelerates. Global frameworks now influence each other in real time, what happens in Brussels reverberates in Washington, Singapore, and São Paulo. In such a world, waiting to react is no longer responsible. It is risky.

Foresight becomes the new currency of trust. Not just seeing change but anticipating its implications before it crystallises. Regulatory intelligence, when embedded in strategy, elevates compliance from a retrospective action to a proactive, business-enabling force.

4.4.2 From Monitoring to Meaning: Tools that Make Sense of Complexity

The challenge is not a lack of information. Organisations are flooded with updates, alerts, draft guidance, and evolving standards. The real challenge is meaning-making. Which signals matter? Which changes are truly material to your product portfolio? What will a minor wording shift in EU MDCG guidance mean for a high-risk software-driven device?

Modern regulatory intelligence platforms ingest updates from dozens of sources, tagging them by geography, product class, and impact area. But the differentiator is not the platform, it is the integration. High-performing organisations connect these tools to product pipelines, design controls, risk registers, and change management workflows. A shift in cybersecurity labelling requirements in Japan automatically triggers a review of labelling SOPs for all connected devices in scope. That intelligence becomes action.

Beyond tools, visualisation matters. Dashboards that map regulatory developments against pipeline timelines, approval cycles, and known bottlenecks transform foresight from abstract concern to operational guidance. Predictive models can forecast regions most likely to introduce scrutiny based on current patterns, enabling teams to pre-adjust documentation or gather additional evidence in time.

The Global Regulatory Foundation in the iQSM is the natural home for this capability. It functions as the early warning system, the interpreter of external change into internal readiness. It makes foresight actionable.

4.4.3 Foresight as a Cultural Habit

No matter how advanced the platform, it is culture that turns foresight into capability. In many organisations, regulatory awareness is still siloed. Updates are noted by specialists but not shared in time. Impacts are surfaced late, once they have already affected design, submission, or approval. The organisations that lead do something very different: they democratise foresight.

In these cultures, regulatory insight is a shared language. Product councils include regulatory foresight as a standing item. R&D sprints begin with briefings on evolving global considerations. Clinical strategy incorporates emerging evidence expectations from Notified Bodies and FDA workshops. Foresight becomes embedded in the rhythm of execution, not as a compliance exercise, but as an enabler of confidence.

This habit is supported by mindset. Regulatory change is not feared but expected. It is met not with resistance, but with curiosity. Teams ask not "Do we need to comply?" but "How can this enhance what we are building?" This is the shift from fear-based to foresight-driven regulation.

4.4.4 From Passive to Proactive

In an iQSM-enabled enterprise, foresight is not occasional. It is continuous. It sits at the intersection of strategy and execution, embedded into the living system. Regulatory scenarios are used in portfolio planning. Submission timelines are shaped by anticipated reviewer pressures. Risk-based design controls are informed by near-future guidance trends, not just current mandates.

The most forward-leaning firms now include foresight KPIs in executive dashboards. They track not only how quickly a team reacts to change, but how often they saw it coming. This redefines regulatory maturity, not by the volume of documentation, but by the ability to anticipate.

It also becomes a differentiator in external relationships. Regulators begin to recognise which companies come prepared, which anticipate issues, and which treat policy evolution not as disruption but as partnership. These companies earn trust faster, face fewer questions, and move with greater velocity.

In the iQSM, regulatory foresight is not a feature. It is a design principle. It transforms the Global Regulatory Foundation from a static layer into a living, breathing, sensing organ, one that translates complexity into clarity and change into competitive readiness.

As we move next into the practical expression of harmonisation through real-world case studies, we will see how foresight, intelligence, and cultural maturity together form the invisible infrastructure behind global excellence. Let us now observe this blueprint in action.

4.5 Case Studies: Global Regulatory Synchronisation in Practice

4.5.1 Alignment as a Strategic Competency

The most successful MedTech organisations are not those with the largest regulatory teams or the most sophisticated compliance software. They are those that treat alignment as a strategic competency. These firms do not react to global regulations, they design with them. Harmonisation is not something they hope to achieve post-development, instead it is embedded from the first sketch of a design input through to final submission and beyond.

Such alignment is not theoretical. It manifests in practical capabilities: design controls that satisfy multiple authorities, documentation structures that flex by region without duplicating effort, and teams that speak in the language of equivalence, not exception. Regulatory synchronisation in these firms is not a tactical goal, rather it is a cultural operating rhythm.

4.5.2 Foresight and Flexibility

Consider a mid-sized neurotechnology firm preparing for simultaneous market entry in the European Union and the United States. Instead of allowing each regional regulatory team to operate in silos, the company formed a cross-market regulatory task force from day one. This team co-created the design history file with visibility into both EU MDR and US FDA QMSR expectations. Where discrepancies existed, such as usability thresholds or cybersecurity documentation, the task force aligned proactively, documenting the rationale and accommodating both perspectives within a single dossier structure.

This early alignment prevented rework and avoided costly delays. But more importantly, it gave the development team a clear sense of direction. There was no ambiguity, no last-minute panic. The foresight built into the process gave confidence to internal stakeholders and credibility to external reviewers. The result was a near-simultaneous approval in both regions, achieved not through heroics, but through harmonised design.

4.5.3 Diplomacy in Action

Elsewhere, a large cardiovascular device manufacturer operating across Latin America recognised that formal harmonisation frameworks were limited. Requirements varied widely. Regulatory timelines were opaque. Instead of responding to each country's demands individually, the company took a different approach: regulatory diplomacy.

They embedded regional regulatory liaisons into the organisation, not just to manage submissions, but to build relationships. These professionals spoke the language, understood the regulatory culture, and cultivated trust with national authorities. They did not wait until submission to engage. They met regulators early, shared development plans, and sought informal feedback before official timelines began.

The payoff came when several countries adopted ISO 13485-aligned systems. This firm was one of the first to benefit, gaining approvals with reduced documentation burdens. Their submissions were not just accepted, they were trusted. The regulators remembered the professionalism, the transparency, and the ongoing dialogue. Harmonisation here was not achieved through documentation alone, but through credibility.

4.5.4 The Common Thread: Integration Before Expansion

What unites these organisations is a simple yet powerful principle: integration before expansion. They resist the temptation to pursue rapid geographic scale without first establishing a regulatory architecture that can absorb variation without chaos. Harmonisation is embedded in their governance structures, their design control workflows, and their product development lifecycles. They do not localise reactively. They localise intelligently, building from a harmonised core.

This approach also transforms internal behaviours. Quality teams understand the nuance of regulatory expectations. Regulatory teams are welcomed into innovation sprints. Product managers learn to see regulatory strategy not as a constraint, but as a blueprint for market confidence. These companies build systems where synchronisation is not an afterthought. It is the operating model.

4.5.5 A Future of Purposeful Alignment

These examples are not just success stories. They are signposts toward a future where regulatory alignment is not a hope, it is a habit. As global regulatory frameworks evolve toward convergence, through

IMDRF efforts, digital submission pilots, and shared audit models, organisations that operate with harmonised intelligence will be those that lead.

What these firms understand, and what others must learn, is that harmonisation is not about flattening difference. It is about enabling connection. A harmonised system does not ignore local variation. It absorbs it intelligently, reflects it respectfully, and responds with precision.

As we now prepare to explore the role of digital tools in this transformation, remember this: harmonisation is not the absence of conflict. It is the presence of purpose. It is what turns complex global ecosystems into coherent systems of trust. Let us now turn to the technologies that make this orchestration scalable, sustainable, and real.

4.6 Digital Bridges: Technology's Role in Harmonisation

4.6.1 The Digital Bridge to Global Unity

Technology is no longer a support act in regulatory harmonisation, it is the conductor. Where policies attempt to align intent, digital tools align execution. The fragmented nature of global MedTech regulation cannot be overcome through documentation alone. It requires systems that can think, translate, and respond in real time. Technology is the language through which harmonisation is operationalised.

As regulatory demands grow more dynamic, and evidence becomes more data-driven, organisations must evolve from legacy compliance models to digital ecosystems. These systems do not merely manage checklists, they orchestrate intelligence, embedding global requirements into the very structure of execution.

4.6.2 From Static Documents to Living Intelligence

The traditional world of regulatory compliance has been document-centric. Submission files were built in silos, reviewed in sequence, and often recreated for every new geography. This model, while familiar, is slow, inefficient, and prone to error. In a synchronised environment, documents give way to data: structured, searchable, and situationally aware.

Modern MedTech leaders operate with cloud-based regulatory platforms that enable real-time updates, version control, and linked evidence management. When a regulation changes, the platform does not simply notify the team, it also traces the impact across SOPs, technical files, risk matrices, and training logs. This shift turns static compliance into living intelligence, always ready, always relevant.

4.6.3 The Rise of the Digital Regulatory Twin

Among the most transformative tools emerging in harmonisation is the digital regulatory twin. Much like its counterpart in manufacturing, the digital twin mirrors the lifecycle of a product's compliance across global markets. It provides a real-time model of how requirements, submissions, surveillance activities, and risk profiles interact, allowing regulatory teams to simulate scenarios, forecast delays, and plan with foresight.

These twins are not theoretical. In practice, they integrate data from regulatory change monitoring tools, quality management systems, product development platforms, and clinical databases. They can model how a change in post-market surveillance expectations in Europe affects labelling in Australia, or how an FDA cybersecurity draft guidance impacts risk assessments for devices already in market.

The digital regulatory twin becomes the central nervous system of the global compliance architecture: constantly sensing, adapting, and guiding decisions with precision.

4.6.4 Turning Complexity into Capability

Harmonisation, when supported by intelligent technology, shifts from burden to advantage. Leading firms are no longer overwhelmed by global complexity. Instead, they harness it. They use regulatory intelligence platforms to ingest and classify global updates. They use AI to flag risks in submission narratives. They maintain global traceability through connected documentation libraries. And they create dashboards not only for regulatory teams, but for executives: offering strategic visibility into readiness, impact, and opportunity.

These capabilities are not limited to large enterprises. Modular, scalable platforms allow even emerging MedTech firms to operate with harmonised excellence. Technology is the great equaliser, reducing friction, expanding visibility, and embedding discipline across functions, geographies, and stages of product maturity.

4.6.5 Embracing the Shift to Digital Maturity

Digital maturity is now a prerequisite for regulatory agility. Organisations that remain document-driven will struggle to adapt to evolving regulations, increasing audit scrutiny, and the expectations of global synchrony. But those that embrace harmonised architectures, where digital tools connect quality, regulation, and innovation, will thrive.

This maturity is not achieved by deploying a tool alone. It requires governance, cross-functional engagement, and leadership sponsorship. It demands that technology not sit in isolation but serve as a harmonising layer across the entire iQSM: from the global regulatory foundation to the living QMS and innovation engine.

When regulatory foresight, modular process design, and digital enablement converge, organisations move from reacting to leading. They no longer fear complexity, they orchestrate it.

4.6.6 Closing Insight: Harmonisation as the Global Conductor of Trust

At its core, digital harmonisation is not about efficiency, it is about trust. Trust between companies and regulators. Between global teams and their systems. Between innovators and the patients they serve.

Technology, intelligently applied, builds that trust. It ensures that when a device moves from Europe to Asia to the Americas, it does so with a consistent standard of evidence, traceability, and accountability. It allows organisations to promise more than access; it allows them to promise alignment, safety, and speed, without compromise.

As the iQSM blueprint evolves from concept to capability, technology is the enabler that makes the system breathe. It does not replace human expertise, it elevates it. It frees regulatory professionals from rework and empowers them with insight. It transforms harmonisation from a goal into a lived reality.

This is not just a call to reflect. It is a call to harmonise. Global success demands local understanding and systemic alignment. Ask yourself and your teams bold, practical questions:

- Are your global regulatory strategies coordinated, or reactive and regionally fragmented?
- How well do you translate global regulatory expectations into consistent product and documentation readiness?
- Are you using regulatory intelligence proactively to shape pipeline and market decisions, or simply reacting to changes?
- Where do launch delays or compliance issues trace back to lack of harmonisation in process, ownership, or tools?

Let us now carry this vision forward. In the chapters ahead, we turn to execution: how synchronised governance, agile process tailoring, and adaptive operating models bring iQSM to life every day, on every project, in every geography. Because strategy alone does not change systems. Action does.

CHAPTER 5

LIVING SYSTEMS: EMBEDDING REGULATION INTO A SYNCHRONISED QMS

Regulation, when misunderstood, is often seen as a static weight, something to be complied with, ticked off, and moved beyond. But that mindset belongs to a fading era. In today's dynamic MedTech environment, regulation must be reimagined not as a distant checkpoint, but as a living presence, woven into the everyday rhythms of operations, decision-making, and innovation.

This is the chapter where that shift comes alive.

As global regulatory frameworks evolve with unprecedented speed, propelled by advances in AI, digital health, cybersecurity, sustainability, and new models of evidence generation, organisations are under pressure to shift from lagging compliance to proactive synchronisation. And yet, the systems they rely on are often out of sync with the speed and nuance of this new reality. Static SOPs, rigid role structures, brittle QMS platforms and template-based implementations are simply unfit for an environment where change is constant and trust must be earned daily.

To thrive, organisations must move beyond compliance events and into the territory of living systems: adaptive, contextual, and synchronised across functions. In these systems, regulation is not translated once and locked into procedure. It is continuously sensed, interpreted, and tailored through intelligent process design, risk-based decision-making and learning culture. It is not delegated to one function but distributed across every role and responsibility. It becomes a presence, not just a requirement.

At the heart of this transformation lies the iQSM. And at the centre of iQSM lies its second foundational layer: the Living Quality System Framework.

This is not a stack of procedures. It is the operational intelligence centre of the entire model. It is where the intent of global regulation, including the frameworks we explored in Chapter 5, is synchronised into executable excellence. It is the synchroniser, the translator, the connective membrane between strategy and action. It listens upward to the global regulatory foundation and downward to frontline execution. It transforms complexity into clarity, rules into behaviour, feedback into foresight.

Within this Living QMS layer, every process, role, document, and decision reflects a harmonised logic, one that allows for precision without rigidity and flexibility without fragmentation. This is where tailored blueprints emerge, processes that flex intelligently by product class, geography, or lifecycle stage, yet remain connected to one global rhythm of quality.

And in this layer, technology is not an overlay, it is embedded infrastructure. Traceability dashboards, intelligent document systems, digital twins, and regulatory insight engines form the nervous system of a synchronised enterprise. These tools do not merely record. They sense, adapt, and guide, enabling regulation to live and breathe inside operations with fluidity and speed.

Finally, and most importantly, this chapter explores the cultural conditions that bring the system to life. Because no amount of architecture or technology can replace the mindset, trust, and shared fluency that makes compliance sustainable. When teams feel safe to surface gaps, when quality becomes part of identity, and when regulation is understood not as constraint but as alignment, excellence becomes natural.

> *"In living systems, regulation is not followed.*
> *It is embodied. It breathes with the work."*

This is not about updating a QMS template or tweaking an SOP. It is about redesigning the very architecture of how MedTech works, building systems where global standards and local execution move in harmony. Where tailoring is intentional. Where documentation is alive. Where technology is a harmoniser. And where culture becomes the true engine of trust.

Welcome to the Living QMS Framework.
Where regulation lives.
Where quality leads.
Where synchronisation begins.

5.1 Translating Regulation into Process

5.1.1 From Policy to Practice: The Missing Bridge

One of the most persistent paradoxes in MedTech is this: organisations understand regulation yet still fail to consistently meet it. The failure is rarely intellectual. It is translational. Regulations are reviewed, interpreted, and even admired in principle, but too often they remain suspended above the operational layers where decisions are made, risks are managed, and products are built. The intention is sound. The implementation is brittle.

The assumption that policies naturally cascade into practice is dangerously optimistic. A procedure may be updated, an SOP may be revised, but unless those documents are translated into workflows that resonate with how teams actually operate, compliance becomes decoupled from performance. And when that disconnect is sustained, audit risk rises, quality outcomes weaken, and regulatory gaps emerge not from ignorance, but from misalignment.

This is not a failure of commitment. It is a failure of design. And design must begin with translation.

5.1.2 Regulatory Blueprints: Tailoring with Intelligence

Living systems resolve this disconnect not by rigidly enforcing policies, but by engineering contextual blueprints, adaptable process designs that anchor to regulatory logic while flexing by risk class, geography, and lifecycle stage. These blueprints are not abstract. They are executable frameworks that allow an organisation to say: "This is how a Class III device in the EU should flow. This is how a software-only product in North America flexes. This is how risk justifies difference."

Rather than develop dozens of disconnected SOPs, synchronised systems begin with a harmonised backbone, anchored to ISO 13485, MDSAP, and other global standards, then create purpose-built variants. Each variant is explicitly mapped to both regulatory intent and business context. This approach does not dilute compliance. It strengthens it, because it removes ambiguity, prevents process drift, and ensures that tailoring is governed, not improvised.

5.1.3 The Process is the Message

Every process is a signal. It tells teams what matters, what to prioritise, and how to act when the rules are unclear. When design control workflows are not aligned with risk management, or change control ignores post-market surveillance, the system begins to send conflicting signals. People are forced to rely on intuition rather than structure. And that is where non-compliance hides.

The most resilient organisations architect their systems to speak with one voice. For instance, when regulatory expectations shift, as with new AI/ML guidance from the FDA, a living QMS does not merely issue a memo. It adjusts the impacted blueprints: clinical evaluation processes are updated, design validation templates are refreshed, and training modules are regenerated. The system evolves in real time. The message remains clear.

5.1.4 Living Processes, Not Static Controls

Traditional QMS structures often treat processes as control mechanisms, fixed pathways to reduce variability. But in a synchronised system, processes are dynamic, intelligent frameworks that evolve with both regulatory signals and operational feedback. They are updated not just during annual reviews, but in response to real-world insight.

Consider a post-market surveillance process that integrates real-time complaint data. As new patterns emerge, the process itself evolves, new checkpoints are added, risk thresholds are recalibrated, and surveillance frequency adjusts without requiring manual intervention. This is not agility at the cost of compliance, rather it is compliance made agile.

5.1.5 The Shift from Paper to Pulse

What distinguishes a living QMS is not the volume of its documentation, but the vitality of its execution. In static systems, quality is proven retrospectively through audits, CAPA responses, and backdated evidence. In living systems, quality is proven in real time, visible in dashboards, traceable in workflows, and auditable by design.

This shift from paper to pulse changes how compliance is experienced. Instead of fear, there is fluency. Instead of surprise, there is readiness. And instead of firefighting, there is foresight. Processes are not just executed. They are understood. They are trusted. They are lived.

As we now explore the foundational components that support these processes, documents, roles, and trainings, we will see how a synchronised system is not just about doing things right, but about ensuring that everyone knows how, why, and when to do the right thing.

5.2 Synchronising Documents, Roles, and Training for Clarity

5.2.1 The Invisible Architecture of Trust

In every quality system, there exists an invisible architecture, composed not of buildings or blueprints, but of documents, defined roles, and the training that links them. These are the fibres through which the system breathes. They are not glamorous, but they are essential. When misaligned, they create friction and fragility. When synchronised, they generate trust, clarity, and confidence at scale.

In legacy systems, these components are treated as endpoints. A document is approved. A role is defined. A training is completed. But in living systems, none of these elements stand still. They evolve, constantly, to reflect the shifting reality of regulations, products, and people. In doing so, they move from static compliance tools to strategic assets.

5.2.2 From Static Documents to Living Knowledge

Documentation is the memory of a quality system. Yet in many organisations, that memory is outdated or fragmented. SOPs, work instructions, and quality plans live in silos, archived for audits, disconnected from action. Updates are reactive, often triggered only by audit findings or regulatory inspections. This approach breeds latency, uncertainty, and risk.

In synchronised systems, documentation is alive. Procedures update as regulations shift. Work instructions adapt to new design methodologies or risk signals. Guidance documents reflect the lived wisdom of teams, not just the requirements of regulators. This living knowledge is powered by traceability, digitally mapped from regulation to policy to procedure to training, ensuring every change has a ripple effect that informs action.

When documents are synchronised across product families, geographies, and lifecycle stages, teams operate with shared language.

Confusion recedes. Execution strengthens. And audit readiness becomes a by-product of clarity, not of scrambling.

5.2.3 Role Clarity in a Dynamic Landscape

Roles within MedTech organisations are increasingly fluid. Cross-functional collaboration is essential, yet without role clarity, collaboration quickly devolves into diffusion. When a risk file changes, who updates the design inputs? When a clinical finding challenges a label claim, who owns the response? These moments reveal not just operational weakness, but structural ambiguity.

Living systems define roles not only by function, but by impact. They create dynamic accountability maps where ownership is clear, responsibilities are traceable, and collaboration is intentional. For example, in a synchronised risk management process, the regulatory lead does not simply advise on compliance; they co-own mitigation planning with engineering and coordinate clinical alignment with medical affairs. Each role is tuned to the purpose of the process, not just the job title.

This clarity becomes especially critical when operating globally. Distributed teams, multiple product lines and regulatory diversity demand a system that can scale accountability without creating confusion. Synchronised role architecture ensures that quality is not someone's job, it is everyone's shared responsibility.

5.2.4 Training as Synchronisation, Not Obligation

Training is often the most underleveraged asset in the quality arsenal. Too frequently, it is reduced to an obligation: complete a module, pass a test, tick a box. But that model assumes knowledge is fixed, context is stable, and people learn in isolation. None of that is true in a living system.

In synchronised organisations, training becomes a mechanism for alignment. It connects evolving regulatory expectations to frontline decision-making. It teaches not only what to do, but why it matters. And most importantly, it prepares people to think, not just to comply.

This is achieved through immersive learning experiences: scenario-based modules rooted in real product cases, simulations that explore regulatory dilemmas, and learning analytics that adapt training to individual performance. When a process changes, retraining is triggered by system intelligence, not by manual intervention. When a deviation arises, training gaps are automatically analysed and resolved.

In this model, training becomes a source of culture, not just compliance. It builds muscle memory for excellence.

5.2.5 Synchronisation as an Organisational Signature

When documents, roles, and training are truly synchronised, the organisation begins to speak with one voice. Not because everyone is doing exactly the same thing, but because everyone is guided by the same rhythm, the same values, the same intent. That is the signature of a living system: aligned, adaptive, and trusted from the inside out.

And this synchronisation is not accidental. It is designed. It is maintained. It is lived. As we move next into the terrain of tailored system design, where global coherence must meet local flexibility, we will see how living systems allow for variation, not by diluting standards, but by mastering them.

5.3 Tailored, Not Fragmented: Customising Without Compromise

5.3.1 The Paradox of Precision and Flexibility

At the heart of modern MedTech quality lies a paradox: the need for rigorous standardisation across global operations, and the equal need for contextual flexibility across products, geographies and risk profiles. Many organisations fail not because they lack one or the other, but because they cannot reconcile both. They attempt to scale with rigid uniformity and end up brittle. Or they decentralise for agility and descend into chaos. The true solution lies not in choosing one extreme, but in mastering the tension between them.

Living systems solve this by designing for purposeful tailoring. The goal is not to enforce sameness. It is to embed clarity into variation, so that every deviation from the global norm is justified, governed, and harmonised. This is not dilution. It is design intelligence.

5.3.2 Blueprints That Flex With Purpose

In a synchronised QMS, flexibility is not a wildcard. It is a structured option embedded in scalable blueprints. These are not static templates, but configurable architectures, process models with predefined anchors, rule sets, and customisation parameters. They enable variation where needed, while preserving regulatory coherence and operational control.

For example, a global design control blueprint may include branching logic based on product risk classification. A Class III implant will follow deep clinical validation and risk scrutiny. A Class I device may follow a lighter route. The architecture is shared, but the execution is tailored. Similarly, localisation logic might be embedded into the change control process, so that country-specific labelling updates follow regionally governed review paths but still originate from a common change trigger.

This is the difference between flexibility and fragmentation. The former is engineered. The latter is accidental.

5.3.3 Contextual Consistency Across Functions and Borders

Synchronisation does not mean enforcing identical workflows in every market. It means ensuring that, regardless of where or by whom a process is executed, the outcomes are equivalent in intent, risk management, and regulatory assurance. That is contextual consistency.

In a harmonised CAPA process, for example, the underlying principles, thorough root cause analysis, systemic evaluation, timeliness, and traceability, are constant. But the tools, languages, review levels, and escalation triggers may vary by geography. A site in Brazil may use different software than one in Switzerland, but both report into the same governance structure, and both are held to the same global performance standards.

This consistency allows global leadership to see patterns, share learnings and drive improvement at scale, without suffocating local initiative. It is a federation, not a factory.

5.3.4 The Discipline to Tailor Without Drift

Tailoring without governance becomes drift. And drift, over time, becomes risk. That is why synchronised systems treat tailoring not as an exception, but as a deliberate strategy. Every customised process or local deviation is documented, justified, and reviewed. Change control tracks the evolution of tailored elements. Training is updated to reflect the unique execution. Audits test both the alignment to local need and adherence to global principles.

This discipline transforms autonomy into accountability. It empowers teams to adapt while holding the line on quality and regulatory expectations. Discipline also signals maturity to regulators; the ability to demonstrate both flexibility and control is now a marker of credibility, especially under frameworks like MDSAP and ISO 13485.

5.3.5 From Fragmentation to Purposeful Flex

When organisations master the logic of tailoring, they unlock resilience. They can scale without stretching their systems to the breaking point. They can innovate in one market and replicate success elsewhere without reinvention. And they can respond to regulatory shifts, whether in Europe, Asia, or North America, without triggering systemic disruption.

Tailoring becomes not a workaround but a hallmark of excellence. It reflects a system that understands its context, honours its complexity, and chooses its adaptations with care.

As we now move into the practical demonstration of these ideas through the iQSM Pillars in Motion, we will see how tailoring, when synchronised, becomes a competitive advantage. Because the true power of living systems lies not just in their design, but in their execution across diverse, demanding, real-world conditions.

5.4 The iQSM in Motion: Pillars at Work

5.4.1 From Framework to Force

A framework gains its power not from elegance on a slide, but from how it performs under pressure. The iQSM, with its carefully structured layers and five synchronisation pillars – Governance, Intelligence, Integration, Flexibility, and Culture – was designed not as theory, but as a force for execution. This section brings that promise to life. It is where alignment becomes movement, where ideas become practice, and where the Quality System Framework proves its strength in dynamic, cross-functional, real-world action.

The iQSM is not just a model for aligning compliance with operations, it is a model for orchestrating performance. It allows global MedTech firms to embed regulation into the bloodstream of innovation, ensuring that quality, speed, and trust move together.

5.4.2 Synchronising Corrective and Preventive Action (CAPA) Across Borders

One of the clearest demonstrations of iQSM in motion is in the global harmonisation of the CAPA process. Historically, CAPA has been a prime source of fragmentation. Sites develop their own root cause templates. Thresholds for escalation vary. Closure timelines depend more on culture than risk. And the result is not just inefficiency, it is also the erosion of learning.

Under the iQSM, the five pillars drive systemic cohesion. Governance defines a global CAPA policy, setting consistent principles while recognising local regulatory nuance. Intelligence allows for proactive signal detection, integrating complaint data, audit findings, and post-market surveillance to identify where issues emerge and where solutions are sticky. Integration ensures that CAPAs do not live in isolation, but are tightly connected to risk files, design history, training plans, and change control. Flexibility allows for local tools and workflows but ensures they all plug into a shared rhythm. And Culture transforms CAPA from a blame game into a collaboration engine, where finding a problem is a badge of integrity, not a source of fear.

In practice, this creates a living CAPA ecosystem. A trend detected in one site becomes a signal for others. Lessons learned in Brazil strengthen practices in Germany. And a unified dashboard allows leadership to see not only how CAPAs are progressing, but whether they are building resilience or simply closing tickets.

5.4.3 Real-Time Learning Loops Between Innovation and Quality

In traditional models, product development and quality systems often operate on separate tracks. Engineering builds, quality ensures compliance. The iQSM rejects this divide. It embeds feedback loops between innovation and QMS execution, turning development programmes into live contributors to system maturity.

Imagine a design team working on a wearable diagnostic device. During clinical feasibility, they uncover usability challenges in elderly populations not fully anticipated in the risk file. Under iQSM, that insight does not remain siloed. The Integration pillar ensures that the finding updates the Human Factors Engineering file, the CAPA system, and even upstream training modules. Intelligence tracks the frequency of similar findings across portfolios, triggering a governance response to strengthen usability protocol reviews earlier in design. And because Culture promotes openness, the team shares the insight across geographies, enabling other programmes to pre-empt similar risks.

This is not a process compliance. It is a process intelligence. It transforms the QMS into a dynamic learning platform, fuelled not just by failures, but by the frontlines of innovation.

5.4.4 Enabling Tailored Governance Through System Feedback

Another often-overlooked benefit of a synchronised system is how feedback from operations shapes policy. In many organisations, governance is top-down: policy is created centrally and pushed out for compliance. In iQSM-enabled environments, the flow is bidirectional.

For instance, through the Governance and Flexibility pillars, a regional operations team facing unique cybersecurity expectations under evolving Asia-Pacific regulations can request policy refinement to reflect their reality. This input, supported by Intelligence from regulatory tracking tools, flows into a governance forum, where the change is reviewed, approved, and incorporated into the next policy update. Training is automatically triggered. Document control reflects the change. Local reality is not just tolerated, it is integrated.

In this way, the QMS becomes not only synchronised across time zones but synchronised across insights. The system gets smarter, not just bigger.

5.4.5 Scaling Trust Through Synchronised Maturity

What unites these examples (CAPA harmonisation, innovation feedback loops, dynamic governance) is the principle of synchronised maturity. It is not enough for one function, one site, or one process to be excellent. In a living system, maturity must be distributed. Every part of the organisation must share a common language of risk, quality, and regulation, even as they apply it in contextually different ways.

This is the final power of the iQSM: it scales trust. Not trust as a vague ideal but trust as operational reliability. A regulatory submission built on processes that have been synchronised across functions and geographies carries more weight. An audit response shaped by harmonised intelligence is more persuasive. A product that reflects learning from global teams earns market confidence more quickly.

> *"Synchronisation is not about perfection. It is about alignment in motion, where people, processes, and principles move with the same intent, even across difference."*

As we now prepare to explore how technology sustains and scales these living rhythms, remember: systems do not become synchronised through control alone. They become synchronised through connection. Through listening, learning, and leading, together.

5.5 Tech Tools for Adaptive Control and Insight

5.5.1 The Shift from Digital Storage to Digital Intelligence

There was a time when the digital transformation of quality systems was celebrated simply for replacing paper. Procedures stored in PDFs, training logged in electronic systems, deviations tracked through structured forms – this was progress. But that era has passed. In the age of real-time regulation, accelerated innovation, and global operations, digital storage is not enough. What organisations need now is digital intelligence.

A synchronised, living QMS is not just digitised. It is responsive. It senses change, flags risk, supports decision-making, and learns over time. Technology in this context is not merely an enabler, it is a co-pilot, amplifying human judgment and aligning systems toward shared purpose. The most forward-thinking MedTech firms no longer ask what their tools can record. They ask what their systems can foresee, synchronise, and strengthen.

5.5.2 Electronic Quality Management Systems (e-QMS): Necessary But Not Transformational

e-QMS remain foundational. They ensure version control, enforce workflows, maintain audit trails, and streamline approvals. But while these platforms provide structure, they do not automatically enable synchronisation. An e-QMS can be rigid. It can replicate siloed thinking in digital form. And it can fail to deliver value if used merely as a repository.

To become transformational, e-QMS platforms must be embedded within a broader architecture that connects rather than isolates. The iQSM model provides this architecture. Through the Integration pillar, e-QMS connects to upstream and downstream systems, Product Lifecycle Management (PLM), Regulatory Information Management (RIM), supplier databases, and post-market surveillance tools, creating traceability across the full product lifecycle. Through the Intelligence pillar, data is not just captured but analysed, transforming routine documentation into live insight.

It is this integration that breathes life into the system. A deviation logged in one site prompts an automated check for related issues across similar products. A change in regulatory guidance triggers updates to SOPs, training assignments, and risk registers. The QMS does not just manage compliance. It anticipates it.

5.5.3 Product Lifecycle Management (PLM) as a Synchronisation Engine

PLM systems serve as the connective tissue between product design, quality, regulation, and operations. When synchronised with an e-QMS, PLM becomes a powerful engine for decision intelligence. It links requirements to verification outcomes, change controls to market submissions, and design history to audit readiness.

Imagine a design change initiated due to a component shortage. In a static system, this might require a series of manual updates across documents, teams, and tools. In a living system, the change automatically triggers a risk assessment update, a training impact analysis, and a review of submission dependencies. Teams are alerted not just to the fact of the change, but to its ripple effects.

This level of orchestration turns quality from a gating function into a strategic asset. It enables product decisions to be made faster, with more confidence, and with full regulatory traceability.

5.5.4 Dashboards: Seeing the Pulse of the System

One of the most significant shifts in digital maturity is the movement from static reports to living dashboards. Properly designed dashboards are not cosmetic. They are instruments of control and clarity. In a synchronised system, dashboards allow leaders to see the pulse of the organisation: open CAPAs, training gaps, audit findings, change control delays, post-market signals, all in one view.

The power of these tools lies not only in visibility but in alignment. When a global MedTech firm operates across 15 markets with dozens of product families and regulatory pathways, a harmonised dashboard ensures that everyone, from the shop floor to the boardroom, is reading from the same page. It allows conversations to move from anecdote to data, from assumption to action.

And most critically, it enables proactive governance. Leaders can identify systemic risks before they surface in audits. They can prioritise resources, course-correct early, and build confidence with regulators by showing a living record of performance and improvement.

5.5.5 The Role of AI in Regulatory Insight and Control

Artificial Intelligence is no longer at the fringe of regulatory systems. It is at the core of their future. Natural Language Processing tools now scan global regulatory updates, extracting relevant changes and proposing SOP revisions. Machine learning models analyse complaint data and detect patterns that suggest emerging product issues. AI-enabled document comparison tools identify inconsistencies in submissions. And predictive algorithms help assess risk based on historical performance, geography, and product type.

Perhaps most transformative is the rise of the digital regulatory twin, a dynamic model of a product's regulatory landscape across all markets. This twin tracks submission statuses, identifies gaps in data, flags dependency risks, and simulates the impact of design changes. It becomes a virtual rehearsal space where teams can test decisions before implementing them in reality.

But AI's value lies not in automation for its own sake. It lies in what it frees humans to do: to think more strategically, to act with more foresight, and to lead with more clarity. The goal is not to remove human oversight, but it is to elevate it.

5.5.6 Synchronised Technology as a Living Backbone

When these tools come together, e-QMS, PLM, dashboards, and AI engines, they form not just a digital ecosystem, but a synchronised operational backbone. A backbone that supports not only quality and compliance, but agility, resilience, and innovation. It transforms regulation from something organisations respond to, into something they anticipate and align around.

And it makes living compliance possible. A complaint triggers not just an investigation, but a system-wide risk assessment. A regulatory update triggers retraining across affected roles. A product issue in one market prompts design improvements everywhere. This is not control by bureaucracy. This is control by insight.

"Technology in living systems does not impose discipline. It animates it. It allows the system to learn faster than the pace of change."

As we now prepare to enter the final section of this chapter, we must acknowledge that none of these matters – no dashboard, no workflow, no AI tool – without the human element. Synchronisation is not sustained by systems alone. It is sustained by people who believe in the purpose of the system, who act with confidence, and who understand that regulation is not a wall, but a window. It is time to turn toward the culture that makes it all possible.

5.6 Living Compliance: Replacing Fear with Fluency

5.6.1 From Policing to Partnership

In many MedTech organisations, compliance still carries the weight of fear. It is seen as an obligation to survive audits, a checklist to satisfy regulators, or a set of rules enforced from above. This mindset, rooted in a legacy of inspection-based oversight, creates distance between quality systems and the people they are meant to support. It fosters anxiety instead of ownership, silence instead of signal, and minimalism instead of excellence.

Living compliance reimagines the relationship. It moves from policing to partnership. It treats regulation not as a hammer, but as a compass. Instead of asking "Are we compliant?" it asks, "How do we use compliance to build trust, accelerate innovation, and deliver safer outcomes?" In this world, the quality system becomes a place of clarity, not confusion; a foundation for informed decision-making, not a constraint on progress.

This transformation is not rhetorical. It is cultural. And it begins with the safety to speak.

5.6.2 Psychological Safety as the Engine of Systemic Integrity

No living system can function if people are afraid to surface problems. When fear dominates, deviations are hidden, issues are rationalised, and audits become exercises in fiction. But when psychological safety is embedded, a different pattern emerges: people raise concerns early, share learnings freely, and use the system not as a shield, but as a strength.

Psychological safety does not mean the absence of accountability. It means the presence of trust. Teams trust that raising a quality issue will lead to resolution, not reprisal. They trust that admitting uncertainty will be seen as strength, not weakness. And they trust that the system they are asked to live is built to support success, not to punish deviation.

This trust must be modelled by leadership. When leaders show vulnerability, acknowledge mistakes, and engage with frontline insight, they create the conditions for integrity. The strongest compliance cultures are not built on enforcement. They are built on invitation.

5.6.3 Training as Culture, Not Compliance

Training has long been treated as a transaction: a module assigned, a quiz passed, a signature logged. But this approach teaches memory, not mastery. And it reinforces the idea that compliance is something external, something to pass, rather than something to live.

Living systems approach training as cultural reinforcement. It is how organisations tell their people what matters, what risks are rising, what behaviours are expected, and what decisions are worth pausing for. Scenario-based learning, real-time feedback loops, and immersive case studies replace slide decks and checklists. Employees are not trained just to follow a process, they are trained to recognise why the process exists, where it flexes, and when escalation is necessary.

In the most mature systems, training is not once a year. It is embedded into the flow of work, triggered by changes in regulation, feedback from audits, or insight from the field. It is seen not as a requirement, but as a right.

5.6.4 Language, Tone, and the Social Contract of Compliance

The way organisations speak about compliance shapes how people behave. When internal guidance documents are filled with punitive language, when SOPs focus more on error avoidance than problem-solving, or when quality reviews feel like interrogations rather than conversations, compliance becomes feared. And fear always drives concealment.

Living systems change the tone. They use language that empowers. Guidance is framed as support. SOPs are written for usability, not for lawyers. Root cause analysis focuses on system gaps, not individual failings. Leaders do not demand perfection. They encourage transparency, humility, and alignment.

This shift may seem subtle, but its impact is profound. It changes the social contract. Employees begin to see compliance not as a burden, but as a shared value. One that protects not just patients, but the integrity of the entire organisation.

5.6.5 Culture as the Ultimate Control System

All quality systems, no matter how advanced their technology or elegant their design, ultimately succeed or fail based on culture. It is culture that determines whether someone speaks up or stays quiet. Whether a CAPA is closed to meet a deadline or resolved to remove a risk. Whether a deviation is corrected or truly learned from.

In living systems, culture is not a soft concept. It is the ultimate control mechanism. It ensures alignment when procedures lag behind change. It safeguards performance when audits are months away. And it sustains excellence when complexity rises.

This culture is built one interaction at a time. In onboarding sessions that connect new hires to the mission of quality. In project reviews that frame regulatory risk as a strategic concern. In dashboards that highlight not just compliance rates, but stories of improvement. And in leadership behaviours that reward curiosity, reinforce integrity, and celebrate the courage to care.

5.6.6 Compliance as a Language, Not a Checklist

The final evolution of living compliance is this: it becomes fluent. It is not something people refer to only in audits or during training. It is something they speak every day, in the decisions they make, the trade-offs they weigh, in the risks they surface, and in the insights they share.

In fluent organisations, quality is not seen as separate. It is embedded in engineering, in supply chain, in marketing, in clinical affairs. It is not measured only by findings. It is evidenced by foresight. And it is not owned only by Quality. It is owned by all.

This is the future the iQSM makes possible: a synchronised system where regulation is no longer feared but understood. Where compliance is not a barrier, but a pathway. And where quality is not just a department, but a discipline, a culture, and a competitive advantage.

This is not just a call to reflect. It is a call to embed. Regulation isn't a checkpoint, rather it is a living part of your system. Ask yourself and your teams bold, practical questions:

- Are your regulatory requirements deeply embedded into daily workflows, or manually translated and bolted-on after the fact?
- How well are your documents, roles, and training aligned to support clarity and accountability across functions?
- Where has "customisation" created unnecessary fragmentation or complexity in your QMS?
- Does your organisation treat compliance as a fluency to build, or a fear to manage?

As we now transition to the Innovation Execution Engine, the third and final layer of iQSM, we bring all of this to life. Because the true test of a living system is not in its design, but in its ability to deliver excellence at the pace of modern MedTech. In the next chapter, we explore how innovation becomes not just faster, but safer. Not just agile but assured. And how, when built on living compliance, execution becomes the most powerful expression of trust.

CHAPTER 6

THE INNOVATION EXECUTION ENGINE: WHERE STRATEGY BECOMES DAILY WORK

S trategy without execution is merely aspiration. In the MedTech industry, where the stakes are measured in human lives and the pace of innovation accelerates daily, the gap between bold vision and grounded delivery is not a luxury to resolve, it is a liability to eliminate. We can no longer afford to operate in models where regulatory compliance, quality discipline, and innovation excellence move in parallel but disconnected lanes. The future belongs to those who can execute in synchrony, with speed, safety, and shared intent.

This is the domain of the Innovation Execution Engine, the foundational layer of the iQSM. It is not just the end of the system, rather it is its proof. The engine is where strategic ambition becomes operational rhythm, where the structure of quality and the logic of regulation are embedded not as constraints, but as catalysts. It is not another process overlay. It is the daily pulse of an intelligent enterprise.

The Innovation Execution Engine embodies the principles of structured agility. It enables fast iteration without fragility, global scale without fragmentation, and compliance without compromise.

Here, execution is not reactive. It is anticipatory. It is not a separate stage. It is the synchronised heartbeat of the system.

This chapter explores how this foundational layer closes the execution gap by transforming frameworks into feedback loops, processes into pulses, and vision into value. It shows how execution, when designed as a tailored layer of the iQSM, connects innovation to regulation not as an afterthought, but from inception. It explains how modular design, contextual logic, and systems thinking coalesce into precision. And most importantly, it reveals how high-performing MedTech organisations turn execution into a differentiator, not by managing complexity, but by synchronising it.

Let us now step into the engine room of iQSM, where execution is not just a phase. It is a force. Where systems breathe. Where strategies move. And where innovation becomes care, delivered with speed, trust, and intention.

6.1 From Vision to Velocity: Closing the Execution Gap

In the corridors of strategic intent, vision often gleams, crafted with care, elevated by ambition, and articulated with precision. Yet in MedTech, vision alone does not deliver impact. The missing layer is execution: the tailored, synchronised rhythm that transforms aspiration into outcomes safely, predictably, and at pace. This section explores the architecture of that rhythm, how high-performing organisations convert insight into intent, and intent into action, without losing fidelity, trust, or control.

"Vision without execution is philosophy. In MedTech,
execution is care made real — safe, timely, and trusted."

6.1.1 The Execution Gap in MedTech

Across the MedTech landscape, a familiar paradox plays out. Organisations set bold goals, invest in innovation strategies, and map out advanced product roadmaps. Yet too often, projects stall, not because the vision lacks strength, but because the systems of execution lack coherence. Innovation races ahead while quality processes lag. Regulatory expectations evolve but are absorbed too slowly. Functional silos misalign priorities, timelines, and interpretations. And what begins with promise ends in costly remediation, rework, or delay.

This is the execution gap. It is not a failure of leadership, effort, or intelligence. It is a system-level breakdown where ambition is not structurally supported. Most critically, it stems from a legacy mindset that strategy is crafted at the top and trickles down through process. In a synchronised system, that model is obsolete. Execution is no longer a downstream function, it is a structural force designed from the outset.

The Innovation Execution Engine closes this gap by translating the strategy layer of iQSM directly into operational behaviour. It builds a tailored execution layer, modular, responsive, and intelligent, designed to scale vision without losing touch with regulatory integrity or clinical need.

6.1.2 Synchronised Execution in a Demanding World

The MedTech industry does not operate in abstraction. It is governed by regulation, driven by evidence, and judged by outcomes. The challenge is not to innovate faster, but to execute faster without compromising safety or coherence. This is where synchronised execution becomes essential.

In this model, product requirements are not isolated technical specifications; they are co-developed with regulatory foresight and quality architecture. Risk management is not an after-the-fact compliance artefact, it is a core design lens applied from the first sprint. Clinical evidence planning is connected with usability and post-market surveillance. The execution system becomes a network, not of departments, but of shared intent, where friction is reduced and learning flows.

Synchronisation replaces linearity. Instead of handoffs, there are conversations. Instead of checklists, there are co-owned responsibilities. Instead of phase-gated delays, there are momentum-maintaining pulses. In this engine, every part of the organisation moves with contextual clarity, not just compliance.

6.1.3 The Velocity That Emerges from Precision

True execution velocity does not come from moving faster, it comes from removing friction. From designing clarity into process. From embedding decision logic where decisions are made. From using modular workflows that scale across risk classes and geographies while remaining harmonised in their intent.

This is the promise of the iQSM's tailored execution layer. It does not impose a one-size-fits-all approach. It provides scalable execution patterns designed to match product class, regulatory environment, and innovation maturity. The system flexes without fragmenting. It enables acceleration without exposure. And it allows teams to operate with rhythm, not panic.

This is not only a technical design, but also a cultural one. When execution is synchronised, confidence rises. Teams do not scramble, they anticipate. Leaders do not manage exceptions, they steward alignment. And patients do not receive delayed interventions, they receive timely, trusted solutions.

Vision without execution is philosophy. In MedTech, execution is care made real: reliable, safe, and synchronised.

As we now move into the architecture of this execution engine, we will explore how tailored design meets global discipline. How movement is made intelligent. And how this final, foundational layer of iQSM transforms potential into precision: one decision, one process, one synchronised rhythm at a time.

6.2 Engineering the Innovation Execution Engine

The heartbeat of a high-performing MedTech organisation is not its strategic plan, but its capacity to execute that plan in the face of uncertainty, complexity, and change. The Innovation Execution Engine is not an abstract idea, it is a living system designed to translate ambition into action and intent into implementation. It is the foundational layer of the iQSM model, where synchronisation is not an outcome but a built-in property. Here, the tailored execution layer takes form: adaptive, modular, and resilient by design.

6.2.1 The Circulatory Design of Execution

Think of the MedTech enterprise as a living organism. Strategy functions as the mind, governance as the skeleton, and execution as the circulatory system. Without this engine, nothing moves: no signal, no insight, no decision. It carries energy, intent, and direction throughout the organisational body. But it must do more than move. It must respond. A true Innovation Execution Engine does not push activity blindly, it modulates it. It accelerates in moments of urgency and decelerates to manage complexity or regulatory risk.

This requires more than functional workflows. It demands a synchronised framework where tailored execution blueprints allow for consistency across the enterprise while accounting for differences in risk classification, product maturity, and geographic nuance. For example, a Class III device entering multiple global markets requires deeper regulatory integration, longer test cycles, and extended design verification. A wellness application intended for a single jurisdiction may follow a leaner blueprint. The structure flexes, but the fidelity remains.

6.2.2 Coordination and Feedback in Real Time

Execution is not just a movement, it is a conversation. In a healthy execution engine, functions are not isolated, they are continuously exchanging information. R&D must not only design a solution but design it in concert with evolving regulatory standards. Quality does not inspect outcomes, as it helps shape them in real time. Post-market feedback is not stored, it is streamed, understood, and looped directly into change control, design updates, and clinical planning.

This level of coordination demands tools and rituals. Integrated PLM and e-QMS platforms provide the digital architecture, while cross-functional sprints, digital traceability maps, and scenario simulations provide the daily pulse. A cybersecurity update does not wait for a change control committee, it triggers a pre-defined action protocol across development and risk management. A usability issue raised in a field study is captured by vigilance tools and piped directly to the relevant design owners within hours.

This is not efficiency for its own sake, it is safety at speed. When execution is designed to synchronise, response becomes proactive. Delay becomes exception. And decisions gain velocity without losing integrity.

6.2.3 Execution as a Living System

The hallmark of the Innovation Execution Engine is that it evolves. Like a living system, it senses its environment. It adapts to regulatory flux, innovation feedback, and post-market complexity. It integrates learning. It builds resilience through action. This means the engine is not a set of static workflows, it is a designed capability, one that gains strength over time.

But no system sustains itself without culture. A living execution engine depends on disciplined teams, psychological safety, and shared ownership. It requires project managers who understand regulatory inflection points, engineers who appreciate downstream risk, and leaders who prioritise clarity over control. Synchronisation is not a checklist. It is a collective mindset.

This is also where digital tools amplify intent. Tailoring logic is not stored in people's heads, it is embedded into process configuration, risk triggers, and dashboards. Alerts are not generic, they are contextual, directed to the person with the decision rights, at the right time. The execution engine becomes both orchestrator and teacher.

**"Organisations do not rise to the level
of their ambition. They rise or fall to
the precision of their execution."**

As we now move deeper into this system, we explore how modularity and flexibility allow execution to scale without fragmenting. Because excellence in MedTech does not lie in being fast, it lies in being fast with purpose, with coherence, and with trust.

6.3 Flexible by Design, Cohesive by Default

At the heart of the Innovation Execution Engine lies a critical design challenge: how to remain scalable without sacrificing precision, and how to allow for local flexibility without compromising global coherence. In a synchronised system, this is not a compromise, it is a principle. Flexibility is built into the architecture, not added as an exception. And cohesion is preserved not through uniformity, but through intelligently governed variation. This is where the tailored execution layer shows its true strength: modular, context-aware, and inherently harmonised.

6.3.1 Contextual Precision, Not Operational Chaos

MedTech organisations serve multiple regions, develop various classes of products, and operate across a range of technologies and maturity levels. It is neither practical nor desirable to impose a single, rigid pathway on every team. Yet, without careful design, flexibility can slip into fragmentation, where each team defines its own path, learns in isolation, and exposes the organisation to compliance risk and inefficiency.

The answer is contextual precision: the ability to adapt execution without losing clarity. For example, a SaMD team working under a fast-paced release model needs agile design controls, lightweight risk assessments, and real-time user feedback loops. A biologics team navigating complex clinical data and global filings needs deep traceability, structured evidence gathering, and robust quality gates. Both are valid. Both must align to a shared core structure.

The tailored execution layer enables this through structured modularity. Each process module – design control, clinical planning, risk management – is anchored in regulatory intent, but adaptable in depth and cadence based on risk classification, intended market, and product maturity. This ensures that every team operates with relevance and rigour, rather than one at the expense of the other.

6.3.2 The Blueprint That Breathes

Rigid templates break under pressure. Intelligent blueprints flex. Within the execution engine, these blueprints act as living guides, driven by selection logic, governed by traceability, and transparent to all stakeholders. They are not mere SOPs, but they are executable frameworks.

Take a design review process. A global framework defines what constitutes a review, who must be involved, and which outputs are expected. However, the cadence, criteria, and tooling differ based on product risk and development model. High-risk programmes might mandate formal reviews with regulatory oversight. Lower-risk, fast-cycle innovations might use peer-led digital checkpoints. The outcomes are still traceable. The intent is still preserved. But the pathway breathes with context.

This is process modularity in action. It allows organisations to move away from binary choices, standard versus tailored, and instead operate in a calibrated space where process is selected, configured, and governed in alignment with product needs and strategic priorities.

> *"A truly global system is not one that treats every problem the same, but one that knows how to make difference coherent."*

6.3.3 Empowering Excellence with Discipline

Tailoring is only powerful when done with discipline. And that discipline comes from a QMS that provides not just procedures, but decision logic. It must make clear which elements are fixed and which are configurable as well as under what conditions tailoring is permitted, who must approve it, and how deviations are documented and integrated into the learning system.

This requires more than documentation, it demands capability. Teams must be trained not only how to follow processes, but how to adapt them responsibly. Cross-functional alignment must be embedded into project planning, with regulatory, quality, and technical leads co-authoring execution pathways rather than retrofitting them after development begins.

Digital tooling supports this with configuration engines that map execution logic to risk profiles, product types, and jurisdictional nuances. Dashboards provide oversight into where and how tailoring is applied, enabling leaders to monitor for drift, enforce consistency, and capture best practices.

This is not bureaucracy, it is governance with agility. It creates a culture where innovation is not constrained by policy but guided by clarity.

6.3.4 A Pathway to Harmonised Agility

When executed with intent, this approach unlocks harmonised agility: the ability to scale operations without losing fidelity, to respond to change without reinventing the system, and to deliver products that are tailored in method but consistent in quality and trust.

This balance is essential. Regulators must see predictability and evidence, not improvisation. Customers must experience reliability, not variability. And teams must operate with confidence, knowing the system supports, rather than hinders, their intent to deliver safely and swiftly.

This is the future of MedTech execution, not rigid adherence, not chaotic freedom, but a tailored discipline where flexibility is not the opposite of control – it is the evidence of mastery.

As we move into the next section, we explore how knowledge itself becomes the fuel of this mastery. Because tailored execution is only sustainable when it learns as it moves, embedding insight, feedback, and reflection into every stage of the journey.

6.4 Knowledge as Fuel: Driving Smarter Innovation

The Innovation Execution Engine does not run on process alone. Its true fuel is knowledge: structured, shared, and continuously renewed. In a system designed for synchronised execution, learning is not a retrospective activity but a daily discipline. It becomes the connective thread between past experience and future performance, allowing teams to innovate not from scratch, but from strength. When knowledge flows freely across functions, geographies, and product lifecycles, execution becomes not only efficient but intelligent.

6.4.1 From Insight to Infrastructure

In many organisations, knowledge is abundant but inaccessible. It sits scattered across systems, buried in outdated reports, or locked inside the minds of experienced staff who carry decades of insight but rarely have the platform to share it. Execution suffers when knowledge is treated as an afterthought or confined to post-mortem reviews. The result is avoidable rework, repeated mistakes, and missed opportunities to connect dots across initiatives.

To transform knowledge into fuel, knowledge must be treated as infrastructure. This means intentional design: knowledge capture baked into every milestone, shared learning forums built into project rhythms, and structured repositories linked to QMS artefacts and execution workflows. Lessons learned become traceable knowledge objects. Risk rationales are not only justified, they are also catalogued and indexed. Regulatory interpretations are preserved alongside product histories, making every decision more transparent and every action more informed.

This is not knowledge management as archiving. It is knowledge architecture as strategy.

6.4.2 Real-Time Reflection, Not Post-Mortem Regret

Synchronised execution requires real-time reflection. Waiting until the end of a project to conduct a formal review is too late in a world where change is constant. Instead, leading organisations embed learning loops directly into the development flow. Retrospectives after sprints. Pause-and-learn checkpoints during key design reviews. Cross-functional signal reviews triggered by new field data or regulatory updates.

For example, a team navigating a change in clinical evaluation requirements captures their interpretation and response into a shared log. This log is linked to templates, training materials, and design documentation, ensuring that the next team facing a similar challenge starts with a foundation rather than a blank slate.

In this model, every deviation, delay, or discovery becomes an asset, not simply resolved, but recycled. The engine learns with each turn.

6.4.3 Making Tacit Knowledge Tangible

Some of the most powerful insights in an organisation are never written down. They live in decision stories, informal advice, and quiet instincts honed through years of experience. This tacit knowledge is often the hardest to capture, yet the most valuable when surfaced.

High-performing systems find ways to make this implicit expertise explicit. Structured interviews with the senior engineer post-project. Scenario-based simulations informed by real-world events. AI tools that detect patterns across design histories, complaint logs, and audit findings, offering guidance based on aggregated intelligence.

Perhaps the most powerful mechanism is cultural. Organisations must celebrate knowledge-sharing as a leadership behaviour, not an administrative task. When reflection is rewarded, when curiosity is modelled, and when questions are welcomed, the walls between knowing and sharing begin to dissolve.

"In a learning organisation, the past
is not archived. It is activated."

6.4.4 A Culture That Learns as It Builds

Sustainable innovation is not just a function of intelligence, it is a function of humility. Organisations that learn as they build are those where failure is analysed, not hidden. Where knowledge is treated as a shared resource, not a personal advantage. Where teams move fast, not because they skip steps, but because they are standing on well-documented foundations.

This cultural shift is what turns knowledge from a static asset into a living capability. It allows junior teams to access wisdom usually locked in experience. It enables rapid onboarding without compromising quality. It builds resilience, not only in systems, but in people.

"Organisations that do not learn must rely on luck.
Those that learn systematically can rely on progress."

6.4.5 Toward a Regenerative System

The goal is not just a smarter team, it is a regenerative enterprise. One that improves not only with each project, but *because* of each project. Where knowledge captured today shortens the learning curve tomorrow. Where insights from post-market data refine design inputs for the next iteration. And where every signal, whether from the field, a dashboard, or a discussion, is integrated into the next cycle of innovation.

In this system, execution is not only synchronised, but also intelligent. The innovation process becomes a spiral, not a line: always moving forward, but always learning. And with this foundation, we are now ready to explore how feedback becomes foresight, how dashboards and pulse checks close the loop and turn data into direction.

6.5 Closing the Loop: Dashboards and Pulse Checks

In high-performing MedTech systems, innovation is not a linear sprint but a continuous cycle. It learns, adjusts, and advances in response to real-world signals. To sustain this rhythm, organisations must build not just execution engines but sensing systems, mechanisms that detect friction before it becomes failure and convert data into direction. This is the role of dashboards and pulse checks. They close the loop between design and delivery, risk and action, intention and outcome.

6.5.1 From Retrospective to Real-Time

Traditional approaches to monitoring rely heavily on retrospective analysis. Quarterly reports. End-of-project reviews. Lagging indicators that confirm what is already known, after the damage has been done. In synchronised execution, this latency is too slow. What is needed is foresight, not hindsight.

Dashboards enable this shift. When embedded within the innovation execution engine, they act as living mirrors, reflecting the real-time status of quality signals, regulatory readiness, project performance, and market conditions. These dashboards are not ornamental scorecards. They are strategic tools, designed for relevance, precision, and accessibility.

A well-constructed dashboard draws from multiple sources, PLM systems, e-QMS platforms, risk registers, clinical databases, and post-market surveillance feeds. It visualises performance in ways that enable immediate response: not just red flags, but trendlines, outliers, and early warnings. Instead of asking "what went wrong?" teams ask, "what needs attention now?"

6.5.2 Digital Twins and Predictive Pulse

Beyond dashboards, leading organisations are adopting digital twins: virtual replicas of systems, products, or processes that simulate how changes will affect outcomes. In MedTech, this means testing regulatory pathways, manufacturing scenarios, or risk mitigations without touching the physical product. It is execution rehearsal, at scale and with precision.

Imagine planning a global product launch. A digital twin models the impact of shifting cybersecurity standards in one market and new clinical evidence requirements in another. It forecasts submission timelines, resource bottlenecks, and potential audit vulnerabilities before a single deviation occurs. Teams move from reacting to preparing.

This predictive capability becomes the execution engine's pulse. It enables leaders to assess not only current performance but future resilience. Pulse checks – short, frequent alignment sessions – then turn this insight into decision. They align quality, regulatory, and innovation leads around shared signals. They reallocate focus. They pre-empt escalation. And in doing so, they maintain synchrony across complexity.

"In a world that moves faster than your governance cycle, your only safeguard is foresight."

6.5.3 From Signal to Strategy

The power of real-time insight is not merely in awareness, but in alignment. Dashboards and pulse checks become strategic instruments when they shape behaviour. When a spike in complaints leads not just to investigation, but to process improvement. When an emerging regulation shifts the prioritisation of design features. When leadership uses data not to interrogate, but to empower.

Closing the loop means making data active. It means shifting from data-rich and insight-poor to insight-rich and action-oriented. And it requires discipline: not just in building the tools, but in using them as part of the daily operating model. Dashboards must not gather dust. Pulse checks must not become rituals of repetition. They must serve purpose, drive focus, and enable responsiveness.

6.5.4 The Loop That Fuels the Future

In a synchronised execution system, closing the loop is not the end of the story. It is the beginning of momentum. The ability to observe, interpret, and respond in real-time transforms execution from a linear effort into a living rhythm. It builds trust, internally and externally. It ensures that strategy remains anchored in reality. And it equips organisations to scale innovation without losing control.

The next generation of MedTech excellence will not be defined by who can design the most novel product. It will be defined by who can sustain precision under pressure. Who can sense when risk shifts, when alignment drifts, and when complexity threatens clarity and responds before impact occurs.

To close the loop is to bring execution full circle. It is to honour learning, empower judgement, and synchronise progress with purpose. And with the loop closed, we are ready to explore how the heartbeat of SHIFT flows through project life, bringing daily synchronisation into the fabric of MedTech execution.

6.6 Synchronised Execution: Bringing SHIFT into Projects

Innovation in MedTech does not move in a straight line. It pulses. From first concept to patient impact, every product journeys through waves of insight, constraint, breakthrough, and revision. But in traditional operating models, these waves often crash in isolation. Quality operates downstream. Regulatory shows up late. Innovation surges ahead only to be pulled back by unmet requirements. The promise of strategy gets lost in the noise of misalignment. SHIFT changes that. When Systems thinking, Harmonisation, Integration, Feedback and Trust are embedded into daily execution, the journey becomes coordinated. Not slower, but smoother. Not controlled but composed. Execution becomes a dance, not a chase.

6.6.1 Design Freeze: The Moment of Intentional Pause

Design freeze is often seen as a checkpoint, a project milestone to be passed. But in a SHIFT-enabled execution engine, it becomes something more: a moment of synchrony. A pause with purpose. A deliberate alignment of all vectors before momentum accelerates further.

At this critical inflection point, cross-functional teams converge. Design teams validate technical feasibility. Quality ensures that design inputs are clear, testable, and traceable. Regulatory leaders map global submission pathways, clarifying dossier implications for Europe, the US, and emerging markets. Clinical and usability insights are integrated, not post-rationalised. Risk controls are confirmed, not assumed. And governance, anchored in shared intent, creates the traceability backbone that will carry the product through manufacturing, launch, and post-market vigilance.

This is not bureaucracy. It is choreography. Design freeze, in this system, is not a lock but a lift-off. Because every function has had a voice, a view, and a verified role. The baton is passed with clarity. The next phase begins with trust.

6.6.2 Post-Market: The Rhythm of Responsiveness

Execution does not end with launch. In fact, it begins again. Because the real world does not care about product timelines. It presents new information – patient behaviours, failure modes, market reactions – that demand attention. In a synchronised system, these signals are not disruptive. They are anticipated.

Post-market teams monitor not just compliance metrics but system health indicators: time to closure on complaints, escalation trends, usability feedback, and emerging regulatory changes. This information flows directly into the execution engine. Not as noise, but as refined input.

A trend in usage errors triggers a usability reassessment. A signal from Health Canada prompts a design mitigation that ripples across other jurisdictions. A near-miss in the field leads to a subtle change in manufacturing specifications, verified through updated design control. None of this requires starting over. Because in a SHIFT-enabled model, the execution system is built to flex, to learn, to adjust, without fracturing.

6.6.3 From System to Culture

What begins as process eventually becomes practice. And what becomes practice, becomes culture. In SHIFT-enabled organisations, execution no longer depends on heroics. It is not a series of escalations and firefights. It is a rhythm. A shared cadence. Teams know when to pause, when to escalate, when to adjust. They are not aligned by force, but by fluency.

Daily stand-ups evolve into synchronisation rituals, not just status updates. Cross-functional reviews become decision accelerators, not performative checkpoints. Roles are clear, not rigid. Tailoring is enabled, not improvised. Documentation becomes a narrative, not a formality.

And most importantly, people feel empowered. Because they understand the logic of the system they operate in. They trust the purpose of the processes they follow. They see the impact of their choices, not only on compliance, but on patients. Execution becomes meaningful.

"In a SHIFT-enabled culture, synchronisation is not something to be enforced. It is something to be lived."

6.6.4 The Execution Engine at Full Pulse

When synchronised execution becomes a daily discipline, the organisation transforms. Product launches become more predictable. Regulatory submissions gain velocity and accuracy. Quality indicators become leading, not lagging. Teams stop chasing alignment and start operating as one.

And through it all, the Innovation Execution Engine hums: quietly, powerfully, continually. It absorbs change without losing control. It learns without losing speed. It scales without losing trust.

This is the future of MedTech operations: not rigid, not chaotic, but alive. Designed with intelligence. Fuelled by feedback. Anchored in trust.

This is not just a call to reflect. It is a call to execute. Ideas alone don't deliver value, synchronised action does. Ask yourself and your teams bold, practical questions:

- Is there a clear bridge between your innovation strategy and how daily work is planned, tracked, and delivered?
- Where are execution gaps slowing your velocity: lack of clarity, coordination, or feedback?
- Are your innovation processes flexible enough to pivot but cohesive enough to scale?
- Do your teams have access to the right knowledge, at the right time, to make informed decisions?
- How consistently do you close the loop with real-time data, dashboards, and project pulse checks?

And as we prepare to explore governance in the next chapter, we now ask: what principles must guide this living system? What structures ensure that it remains ethical, equitable, and human? Synchronised execution delivers velocity. But governance, thoughtfully designed, ensures that velocity stays true. Let us now turn to the compass that guides the system forward.

CHAPTER 7

GOVERNANCE REIMAGINED: FROM STATIC OVERSIGHT TO DYNAMIC ENABLEMENT

In every bold transformation, there comes a point when the very structures built to protect can begin to restrain. In MedTech, that point has arrived. Governance, once the scaffold that ensured quality and safeguarded trust, now finds itself misaligned with the speed, complexity and interdependence of modern innovation. While designed for assurance, many governance models now struggle to keep pace with the dynamic demands of global product development, regulatory evolution, and real-time risk management.

Yet governance is not obsolete. Far from it. It is more essential than ever, provided it evolves. In the iQSM architecture, governance is not a layer among others, it is the dynamic wraparound that ensures synchrony between them. It sees across Global Regulatory Foundation, the Living QMS, and the Innovation Execution Engine, not as discrete silos to inspect but as a continuum to orchestrate. Governance, reimagined, becomes the synchroniser-in-chief: maintaining coherence, enabling trust, and adjusting the rhythm of the system as contexts shift.

This chapter is not about deconstructing governance. It is about redesigning it. Not as static oversight, but as dynamic enablement.

Not as a hierarchy of permission, but as a platform for accountability, transparency, and learning. We will explore how governance becomes distributed but not diluted, how it uses digital intelligence to anticipate rather than react, and how it becomes agile without becoming ambiguous.

In a world where innovation moves faster than review cycles, and where compliance is both global and fluid, governance must adapt. It must become the conductor that senses, guides, and tunes the organisation daily. It must move from being the gatekeeper of control to the guardian of coherence. Because in the iQSM model, strategy is translated into reality not only through execution, but through the governance that gives it alignment, clarity, and course correction.

This is not the end of governance. It is the moment it steps into its highest role yet. Welcome to the orchestration engine of iQSM.

7.1 From Control to Coordination: Governance Reimagined

7.1.1 The Governance Dilemma

Historically, governance in MedTech has served as a bulwark of oversight, a protective construct built to ensure safety, compliance, and continuity. It was grounded in good intention and regulatory necessity. But as speed becomes a proxy for survival, and as cross-functional synchronisation defines success, that construct is revealing its limits. A governance system that once delivered predictability can now breed paralysis. Designed to catch mistakes after the fact, it often lacks the tools to prevent them in motion.

In the iQSM ecosystem, governance cannot remain confined to compliance assurance alone. It must be the active force that aligns the regulatory foundation, the living quality system, and the execution engine into one synchronised rhythm. It is no longer a standalone checkpoint, but a systemic oversight embedded within the flow itself. Governance sees how decisions echo across layers, how risk in design impacts regulatory strategy, and how quality signals influence innovation timelines. It becomes the organisational sense-maker.

7.1.2 From Bureaucracy to Orchestration

The governance of the past depended on hierarchy: vertical escalation, committee-based decisions, long review cycles. These methods worked in linear systems, but they falter in complex, adaptive networks like modern MedTech organisations. Governance today must shift from a passive gatekeeper to an active conductor, guiding synchrony, adjusting cadence, and sensing disharmony before it becomes delay.

This requires a mindset shift from bureaucracy to orchestration. Instead of governing with static rules, we govern with living routines. Instead of enforcing checkpoints, we enable flow. Imagine agile councils replacing static boards and governance that is embedded, not imposed. These cross-functional units don't wait for problems to be reported, instead they proactively surface them in real time through integrated feedback loops and dynamic process tailoring reviews. It is a model where authority follows insight, not seniority.

7.1.3 Governance as the Layer that Sees Across Layers

In the iQSM model, governance operates across all dimensions, not just within a function, but above, around, and between them.

It connects signals from regulatory change management to QMS configuration updates, and from execution bottlenecks to global launch risks. It ensures that tailoring at the product level is traceable to system-level rationale and that digital signals from one function led to strategic course corrections in another.

This is not conceptual, it is operational. For example, governance may oversee a live dashboard showing how a new MDR requirement has triggered a cascade of adaptations across multiple device development programmes. It ensures the tailoring logic applied in quality planning is reflected in training, documentation, and real-world surveillance. In doing so, it ensures that agility does not dilute alignment.

7.1.4 Digital Tools as Governance Accelerators

Digital governance platforms are the nervous system of this reimagined architecture. They do not merely collect data. They connect insight to action. By integrating process tailoring reviews, change control alerts, risk recalibration protocols, and layered performance dashboards, governance is no longer a lagging reviewer. It is a real-time orchestrator.

Consider a scenario where an emerging cybersecurity regulation in Asia triggers alerts across relevant product portfolios. Within a digitised governance environment, impact analysis, risk control updates, and regulatory response plans can be tracked and approved across functions in days, not months. Digital tooling allows governance to be as fast as it is accountable.

7.1.5 Governance as a Cultural Signal

Perhaps the most profound transformation is this: governance becomes a cultural artefact. Not just a system of rules, but a daily expression of clarity, ownership, and integrity. When done right, governance is not a meeting. It is the confidence teams carry when they know decisions are aligned, escalations are fair, and excellence is expected.

This shift does not dilute standards, it elevates them. It creates a space where innovation can move with safety, where agility is bounded by responsibility, and where performance is defined not only by outcomes but by the coherence with which they were achieved.

Governance reimagined is not the pause between planning and action. It is the rhythm that makes both possible. In the iQSM framework, it is not a separate layer. It is the synchronised conductor of every layer.

> **"When governance flows with the rhythm**
> **of the organisation, it ceases to be feared**
> **and becomes deeply trusted."**

7.2 Structure for Synchrony: Balancing Agility and Accountability

7.2.1 Architecture That Moves with Purpose

Agility without accountability is chaos. Equally, control without fluidity is inertia. In high-performing MedTech organisations, the architecture of governance must serve as both the backbone and the nervous system, holding the enterprise upright while enabling dynamic adaptation. Within the iQSM framework, this structure is not layered on top of operations, it is embedded within them. It supports movement, not just mandating it. It creates coherence across layers without constraining individual roles or local execution.

This evolution begins by abandoning rigid, static organisational charts in favour of living structures, ones that flex in real time to accommodate product risk, market variation, and regulatory complexity. Project governance becomes distributed and anticipatory. Authority aligns to outcomes rather than hierarchy. And roles are defined less by title and more by their place in the value stream.

7.2.2 Distributed Authority, Centralised Purpose

A synchronised system does not imply central control. It implies central clarity. Roles must be configured to act with autonomy but remain aligned to a shared intent. In this model, local leaders are not merely executing centrally dictated protocols, they are adapting global frameworks to their context, informed by governance pathways that guide rather than prescribe.

For instance, a regional regulatory lead managing an accelerated programme in Japan may tailor documentation flows based on local requirements, but the tailoring logic is aligned to global governance parameters. A digital trail ensures visibility, and a structured review loop evaluates both local decisions and global coherence. This interplay between freedom and fidelity is the hallmark of modern governance.

7.2.3 Governance Routines as Anchors

In traditional systems, governance was episodic: review meetings, audits, and stage gates that occurred after work was already completed.

In iQSM-aligned organisations, governance becomes rhythmic. It lives in the routines: pulse reviews, agile steering sessions, cross-functional risk walk-throughs, and change impact huddles. These routines operate with defined cadence and purpose. They are not interruptions to the work, instead they are the mechanisms through which the work stays aligned.

Such routines also support tailored execution reviews, ensuring that deviations from standard process (based on risk, geography, or product maturity) remain structured and justifiable. This is not flexibility by exception, it is flexibility by design, governed by rules that are transparent and understood.

7.2.4 Visibility Over Supervision

A fundamental feature of this architecture is that control is replaced with visibility. When teams can see across the system, through digital dashboards, live risk registers, and decision logs, governance becomes a shared responsibility. Leaders intervene when necessary, but more importantly, they enable those closest to the work to operate with clarity and confidence.

This creates a governance model that is no longer driven by escalation, but by enablement. Escalation pathways still exist, but they are not the default. The goal is to resolve issues at source, guided by clear thresholds, cross-functional collaboration, and real-time data.

7.2.5 Adaptive Authority in a Changing Landscape

The regulatory landscape is no longer predictable. Standards evolve rapidly. Expectations shift. Products blur boundaries between software, hardware, and services. To keep pace, governance structures must be capable of self-adjustment. This means dynamic ownership, authority that can shift based on the nature of the decision, the proximity to risk, and the speed required.

In practice, this could look like a programme-level decision council where R&D, quality, and regulatory leads hold rotating responsibility for lead coordination depending on phase and risk profile. Governance is no longer static role fulfilment; it becomes situational leadership, choreographed for synchrony.

7.2.6 Visionary Takeaway

Structure in the iQSM context is not the opposite of agility, it is its enabler. When roles are understood, when review loops are structured, and when clarity is visible at every node, the system moves not with friction, but with flow. This is governance by design, not default. An architecture that shifts from rigid oversight to rhythmic orchestration. One that does not slow innovation but gives it integrity and direction.

7.3 Accountability in Action: Clarity, Ownership, Results

7.3.1 The End of Diffusion

In legacy operating models, accountability often dissolves in the shadows – unclear handovers, vague roles, diffused ownership. The results are predictable: delayed decisions, misaligned priorities, and rising frustration. Not because people lack capability, but because the system lacks clarity. Within the iQSM governance layer, accountability is not a passive outcome of structure. It is a deliberate act of design.

Ownership must be visible, contextual, and consistent. Responsibility must live with the individuals and teams best positioned to act, not as a burden, but as a signal of trust.

Synchronised execution thrives when decision rights are clearly defined, when escalation paths are known in advance, and when accountability is accompanied by the tools, data, and authority needed to deliver outcomes. Ambiguity is not flexibility, but a governance failure.

7.3.2 Radical Clarity Across the System

Clarity is not a by-product of communication. It is an input to high performance. In the most mature systems, governance provides clear mapping between strategic priorities, operational metrics, and individual accountabilities. A regulatory lead knows not only what they are responsible for, but how their decisions impact quality timelines, market access, and innovation velocity. Likewise, innovation teams understand how early-stage technical decisions create downstream compliance implications and who they are accountable to for managing them.

Shared accountability is also embedded in cross-functional milestones. For instance, when design freeze is reached, it is not just an engineering checkpoint. Regulatory, quality, and clinical leads co-own the decision, each with a stake in risk alignment, evidence maturity, and process readiness. This multi-node accountability reduces friction and increases fidelity across layers.

7.3.3 Decision-Making Close to the Work

True accountability empowers those closest to the issue. This means devolving decision rights to where they belong, to the people with the best context and clearest view of the implications. A country-level quality manager navigating an emergent regulatory update should not wait for global approval to act. Instead, the governance system should provide a framework of pre-agreed thresholds, escalation criteria, and audit visibility that empowers safe and timely action.

These "decision permissions" are not informal. They are documented, traceable, and reviewed during regular governance loops. This form of distributed authority does not reduce rigour. It raises the standard for informed autonomy and continuous alignment.

> *"True accountability is not what happens*
> *when something goes wrong. It is what prevents*
> *things from going wrong in the first place."*

7.3.4 Digital Accountability Infrastructure

Technology plays a pivotal role in reinforcing ownership. Governance dashboards track not just tasks, but commitments. Role-based access ensures the right people have the right visibility. Workflow engines route approvals in real time, while audit logs track who made what decision, when, and with what rationale. This digital architecture transforms accountability from a managerial expectation into an operational fact.

More importantly, these systems also surface patterns: recurring delays at specific handoffs, repeated ambiguities in role responsibilities, or bottlenecks in risk escalations. With this insight, governance councils can evolve the system, refining structures and re-aligning expectations before failures compound.

7.3.5 A Culture That Holds and Lifts

Clarity without culture can become command-and-control. But when accountability is embedded in a culture of psychological safety, it becomes a source of pride, not pressure. People are willing to own decisions, raise concerns early, and learn from mistakes because the system rewards transparency and integrity.

Leadership must actively model this ethos. When a senior executive openly discusses a regulatory misstep as a learning moment, complete with system-wide changes to prevent recurrence, they build not only trust but also credibility for the governance model itself.

7.3.6 Visionary Takeaway

In the iQSM governance layer, accountability is no longer abstract. It is mapped, measured, and experienced as clarity in motion. Teams do not waste cycles on second-guessing, nor do they hide behind ambiguity. They step forward, because the system makes ownership safe, visible, and empowering. In this way, governance does not constrain the system, it elevates it. Through clarity comes action. Through action, results.

7.4 Digital Agility: Enabling Insight, Transparency, and Speed

7.4.1 From Static Oversight to Dynamic Enablement

In the traditional paradigm, governance moved at the speed of the reporting cycle. Metrics were collected, analysed, and acted upon retrospectively, often after the opportunity to influence meaningful change had passed. But in today's connected, high-velocity MedTech environment, such latency undermines both compliance and innovation. Governance must now operate in real time, synchronised with product development, regulatory surveillance, and quality performance. This is where digital agility becomes the defining force.

The iQSM model envisions digital infrastructure not as a repository of data, but as a dynamic conductor of decisions. Digital agility empowers the governance layer to sense shifts, respond swiftly and orchestrate alignment across functions and geographies. It converts governance from observation into anticipation, from reporting into enablement.

7.4.2 The Rise of Digital Conductors

Modern governance tools are not dashboards in isolation. They are full systems of orchestration. A digital conductor weaves together information from the tailored execution layer, global regulatory change feeds, quality risk profiles, and innovation roadmaps. It provides leaders with a real-time line of sight across process performance, decision readiness, and risk hotspots.

Consider a scenario where a new cybersecurity requirement emerges in a key market. A traditional system would escalate this issue through weekly meetings and local quality reviews. A digital-first governance model flags it instantly, aligns with impacted product teams, recalibrates risk documentation, and pushes updated guidance through the knowledge layer of the execution engine within hours, not weeks.

This is not theoretical. Leading organisations are already implementing governance hubs built on real-time signal aggregation, role-specific decision boards, and predictive insights from machine learning algorithms trained on historic project delays and quality events.

7.4.3 Real-Time Feedback Loops as Governance Muscle

The digital governance layer must include more than metrics. It must include movement. Feedback loops that are digital, structured, and continuous. These can take the form of fortnightly cross-functional syncs, risk recalibration pulses, or automated compliance scans. Such rituals become the rhythm that keeps the system in motion without losing alignment.

One high-performing MedTech company implemented a "governance heartbeat," a weekly pulse check where leaders reviewed not static dashboards, but live signals drawn from audit trails, process execution status, market shift indicators, and cross-functional sentiment analysis. The conversation was not about past performance, but forward risk: what needs attention, what needs support, and where adaptive decisions are required.

This is what it means to operationalise governance agility.

7.4.4 Integrating the Layers: Vertical and Horizontal Transparency

For governance to truly serve its role within iQSM, it must see across and into the system. That means enabling both vertical transparency, from product team to leadership, and horizontal transparency, across the innovation, quality, and regulatory streams.

Digital platforms must map the interplay of tailoring decisions from the QMS layer, the pulse of innovation in execution, and the signals from real-world use. These connections must be visible, traceable, and contextualised. For example, a deviation in verification protocols during execution should trigger a governance flag, not only within the product team, but at the process performance level, ensuring that QMS tailoring guidelines are evaluated and refined accordingly.

In doing so, governance does not just monitor the system, it improves it. Every insight becomes a design input for evolution.

7.4.5 A Foundation for Informed Trust

Transparency powered by digital agility is not about surveillance. It is about confidence. When leaders can see emerging risk, when teams can align on shared context, and when regulators can observe traceable decision-making, the system earns trust, not through words, but through evidence. The digital layer ensures that governance is neither reactive nor theoretical. It becomes lived. It becomes dependable. It becomes worthy of trust.

This is the moment when governance becomes invisible in the best possible way: it works so well that it feels like intuition. But behind that intuition is a finely-tuned digital nervous system: sensing, learning, and enabling, every day.

> *"The most effective structures do not control the people within them — they align their movement towards a shared intention."*

7.4.6 Visionary Takeaway

In a world of speed, volatility, and complexity, governance must evolve beyond control into clarity. Digital agility enables that shift. It gives governance the power to move with the system, not behind it. It is not just a layer. It is the connective intelligence of iQSM. When designed with purpose, it transforms oversight into foresight and builds a system that leads with both rigour and responsiveness.

7.5 Governance by Design: Making Oversight a Force for Ownership

7.5.1 Governance as an Enabler, Not a Constraint

In the classical model of governance, the term "governance" often conjured images of review boards, corrective action memos, and top-down compliance controls. It served as a necessary defensive mechanism in a risk-averse environment. But in a synchronised system like iQSM, where execution, quality, and innovation are deeply integrated, governance must transcend its legacy as a gatekeeper. It must become an active enabler, equipping individuals and teams with the clarity, confidence, and autonomy to act in alignment with purpose.

Empowered governance is not a relaxed form of oversight. It is a more intelligent one. It is structured with intent but applied with flexibility. It enables teams to move swiftly not in spite of governance, but because of it. When designed well, governance becomes a scaffolding that supports excellence, not a cage that restricts it.

7.5.2 Trust as the Structural Principle

The true power behind empowered governance lies in trust. It is the belief that people, when given the right information and frameworks, will act responsibly. This is not naive optimism, it is operational realism. In high-performing MedTech systems, psychological safety is built directly into governance design. People are not penalised for surfacing issues early; they are rewarded for it. Escalation is seen not as failure, but as foresight. This ethos transforms the cultural tone of governance from fear to flow.

An example comes from a global MedTech manufacturer that reframed its deviation management system. Rather than counting incident volume as a risk indicator, the company used reporting frequency as a signal of system engagement. In doing so, they built a culture where transparency was rewarded, not avoided. The governance system became a support structure for truth, not just a scoreboard for compliance.

7.5.3 Participation That Powers Ownership

Empowered governance also depends on engagement. It must not only be something done to people, but something done with them. In practice, this means designing mechanisms that allow those closest to the work to influence how governance functions. Front-line insights must be fed directly into decision pathways. Role-specific dashboards must show not only what is being tracked but why it matters. Governance rituals, such as review cycles, decision forums, and change control meetings, must include the people impacted, not just those evaluating from afar.

In iQSM, this empowerment is particularly critical. Product teams operating in the tailored execution layer must have a clear understanding of their process responsibilities, tailoring latitude, and regulatory implications. But they must also know how to participate in governance processes that continuously refine those rules. Governance becomes not only oversight, but also co-creation.

7.5.4 From Compliance to Commitment

The shift from static control to dynamic enablement culminates in a redefinition of compliance itself. No longer a check-box to be satisfied, compliance becomes a natural consequence of ownership. Teams do not comply because they are told to, they comply because they understand. Because they helped shape the framework. Because they see how their actions connect to patient outcomes and system trust.

This commitment-based model transforms governance into a culture of stewardship. It demands that leaders model humility, clarity and alignment. It requires systems that are transparent by default, collaborative in design, and generous in insight. And it creates a workplace where decisions are not delayed for permission but accelerated by shared intent.

7.5.5 Visionary Takeaway

Empowered governance does not compromise control. It elevates it. It is not loose, it is lucid. It is not casual, it is confident. It is built on the assumption that people want to do the right thing, and on the reality that when systems trust them to act, they usually do.

This is the heart of governance reimagined: a structure designed for human excellence. When the system trusts its people, the people trust the system. And from that trust emerges not only compliance, but collective brilliance.

7.6 Keeping the Pulse: Governance in Motion

7.6.1 The End of Static Control

Governance in its historical form was designed to endure. It prized permanence, structure, and predictability. This made sense in an era of stable markets and gradual regulatory change. But in today's MedTech landscape, defined by rapid innovation cycles, shifting global standards, and increasing stakeholder scrutiny, static control no longer sustains performance. In fact, such controls can become an organisational liability.

Governance in a synchronised system like iQSM cannot afford to lag. It must evolve from rigid checkpoints to dynamic motion. This means shifting from retrospective oversight to embedded, anticipatory guidance. Rather than governing after-the-fact, leading organisations are building systems where governance breathes with the rhythm of innovation itself: flexing, sensing, and adapting in near real time.

7.6.2 Steering the Ship While It Moves

True governance maturity is revealed not in how a system performs during planning, but in how it responds during movement. A synchronised system is never still. It is alive with decisions, escalations, learning loops, and directional shifts. Governance, therefore, must be designed not to hold the ship still, but to steer while it sails: gracefully, intentionally, and without derailment.

This is achieved through rhythm. High-performing organisations embed frequent and deliberate touchpoints into their operating cadence. These are not bloated review boards. They are nimble alignment forums (weekly operational huddles, rolling scenario-based reviews, cross-functional retrospectives) that replace hierarchy with coherence. Teams course-correct before risk becomes error. Strategy adjusts before lag becomes drift. Governance becomes the act of staying synchronised.

A leading diagnostics firm implemented biweekly governance sprints focused on portfolio alignment. Every two weeks, risk signals, innovation readiness, regulatory shifts, and cross-market implications were assessed collaboratively. These sessions enabled decision-makers to shift resources in real time, pivot regulatory strategies, and respond to emerging evidence before issues hardened into delays. This was not speed at the expense of rigour; it was agility amplified by clarity.

7.6.3 The Rituals That Build Resilience

Governance in motion is not built solely on process, instead it is built on ritual. It is embedded in the way organisations think, gather, decide and learn. When teams come together regularly to surface issues, explore trade-offs, and reinforce principles, a cultural rhythm forms. That rhythm becomes the system's memory and its muscle.

Daily dashboards replace lagging spreadsheets. Weekly decision pulses supplement formal escalation chains. Root cause reviews occur as learning sessions, not post-mortem blame rounds. In such environments, trust deepens, not because everything goes right, but because there is confidence that when things go wrong, the system knows how to respond.

These rituals are what sustain governance when strategy shifts, when market conditions change, or when crises emerge. They are the immune system of a synchronised enterprise, always scanning, always responding, always learning.

7.6.4 From Framework to Dialogue

Perhaps the most powerful evolution of governance in iQSM is its transformation from static framework to living dialogue. It is no longer just the structure that holds the system, it is the voice within it. Governance moves through conversation, through signal, through shared understanding of what matters now and what must happen next.

This kind of governance does not slow innovation, it anchors it. It does not delay decisions, it empowers them. It does not suffocate creativity, it sharpens it through principled guardrails. And most importantly, it enables trust to emerge not from enforcement, but from consistency.

When governance is designed to move with the organisation, not above or behind, it becomes more than a system. It becomes the organisation's conscience and compass.

7.6.5 Visionary Takeaway

In the end, governance reimagined is governance reborn, not as an administrative burden, but as a strategic capability. A dynamic, human-centred, digitally enabled force that protects trust, enables innovation, and ensures that quality is not just maintained, but lived.

It is not static but symphonic. Not episodic but rhythmic. Not feared but followed. And in that motion, in that quiet, purposeful beat, governance becomes what it was always meant to be: the pulse of organisational integrity.

This is not just a call to reflect. It is a call to reimagine. Governance isn't about control, it is about enabling clarity, ownership, and momentum. Ask yourself and your teams bold, practical questions:

- Is your governance structure designed to enable collaboration and speed, or to enforce oversight and control?
- How clearly are roles, responsibilities, and accountability defined, and do they empower or constrain your teams?
- Are your decision-making processes transparent, timely, and trusted across functions?
- Do your governance forums actively balance agility with risk, especially in innovation-heavy programs?
- Is your governance model evolving with the pace of your organisation, or stuck in legacy rhythms?

As we now transition to the next chapter, we shift our focus to trust itself, not as an abstract aspiration, but as the outcome of every decision, process, and relationship. Because in a world accelerating on every front, trust is not optional. It is the currency of progress. And the systems we build must be designed to earn it, every day.

SYNCHRONISED EXCELLENCE IN ACTION – PEOPLE, PROJECTS, & PERFORMANCE

I f Section II built the architectural core, this is where the system comes alive, where synchronisation is tested, refined, and realised in the trenches of daily MedTech work.

This section is about movement. It is where theory becomes practice, structure meets behaviour, and teams transform intentions into outcomes. The SHIFT mindset comes off the whiteboard and into design reviews, sprint meetings, clinical validations, audit rooms, dashboards, and team conversations.

We open with Chapter 8: Trust by Design: Uniting QbD, Risk, and Human Factors for MedTech Excellence, the essential trifecta of patient-centred excellence. In a world where one oversight can jeopardise lives, trust must be designed in, not inspected later. This chapter reframes Quality by Design (QbD), ISO 14971-driven risk management, and human factors engineering not as parallel disciplines, but as a synchronised toolkit for building safe, intuitive, and resilient products from the start. In MedTech, trust isn't earned by ticking boxes, it is engineered by intent.

But the nature of design is evolving. Enter Chapter 9: Digital Lifecycles: Orchestrating Innovation in Software, AI, & Smart Devices. As MedTech converges with Silicon Valley, new lifecycles demand new mindsets. Agile methodologies, continuous deployment, digital twins, and AI are not just tech trends, they are operational challenges that must still meet the rigor of regulatory compliance and patient safety. This chapter explores how to harmonise agile delivery with MedTech responsibility, bridging the once-unbridgeable worlds of software and science.

Yet no technology delivers value in isolation. In Chapter 10: Synchronising Minds: Uniting Cross-Functional Teams for MedTech Momentum, we return to the human engine of MedTech: teams. This chapter dives deep into how cross-functional collaboration becomes more than a meeting invite and how it is built into rituals, incentives, communication tools, and problem-solving mindsets. When quality engineers, regulatory leads, and R&D partners co-create instead of co-exist, speed increases, risk drops, and solutions feel… inevitable.

But how do you measure this new form of excellence? That is the focus of Chapter 11: Metrics that Matter – Predictive Quality & Performance Intelligence. Synchronised organisations don't just report history, they predict trajectories. It is not about more metrics, rather it is about the right ones. This chapter helps you cut the noise, act on what matters, and build a data culture owned by those who use it. When performance intelligence is woven into everyday decisions, you shift from reactive to resilient.

In Chapter 12: Beyond Compliance: Turning Audits, V&V, and Docs into Value Drivers, we challenge the myth that compliance work is a cost centre. Audits become feedback loops. Documentation becomes strategic intelligence. V&V becomes a driver of innovation readiness. Synchronised organisations use compliance to accelerate, not just to protect product success.

Finally, Chapter 13: Simplified Quality: Agile Thinking, Lean Practices, & Smart QMS Tools makes it clear: excellence doesn't have to be complex. This chapter lays out the principles of simplification, empowering teams to do high-quality work without bureaucratic drag. Smart tools, lean thinking, and agile training don't just lighten the load, they sharpen the edge.

Together, these six chapters are your tactical accelerators, the practices and mindsets that turn synchronisation into something visible, measurable, and scalable.

Because in the end, it is not the framework that delivers better MedTech, it is how we live it, lead it, and repeat it.

Let us bring synchronisation to life.

CHAPTER 8

TRUST BY DESIGN: UNITING QBD, RISK, AND HUMAN FACTORS FOR MEDTECH EXCELLENCE

T rust is not a nice-to-have in MedTech, it is the very foundation of every device, system, and decision. Patients may never see a design history file or read a regulatory submission, but they place their lives in the hands of what those processes create. That trust must be earned, designed in from the very start, and protected throughout the product lifecycle.

This chapter explores how to build trust into the products by design. We look at three disciplines: Quality by Design (QbD), risk management, and human factors engineering and show how, when brought together, they create more than safety or compliance. They create confidence.

- QbD ensures we design it right the first time.
- Risk management (ISO 14971 in practice) ensures we identify, evaluate, and control what could go wrong.
- Human factors ensure we design for people, not just specifications.

Individually, these practices add value. Integrated, they become a force multiplier, a living framework for trust.

By the end of this chapter, you will see how trust can be embedded directly into the DNA of products, systems, and culture, making "Trust by Design" not just a principle, but a competitive advantage.

8.1 Setting the Stage: Designing for Trust

Trust is no longer earned at the moment of approval. In the world of medical technology, trust begins at conception, and it must be consciously designed into every detail that follows. It is not a label affixed after validation, but a living principle that governs intent, guides execution, and ultimately defines success.

In today's healthcare landscape, where devices enter homes, diagnose remotely, and integrate with complex digital ecosystems, the margin for ambiguity has vanished. Trust serves as a fundamental element within MedTech. While not always visible, it remains essential and significantly influences the industry. It is not a benefit. It is a precondition.

This trust does not arise spontaneously. It is constructed through thousands of interdependent choices, choices that shape not just compliance, but confidence. From the selection of biocompatible materials to the architecture of software interfaces, every design decision either affirms or erodes the contract we hold with patients, clinicians, regulators, and society at large.

Yet too often, trust is treated as a consequence, not a design principle. Teams chase timelines, scope features, and assume usability, safety, and quality can be appended late, like finishing touches on a nearly completed home. But MedTech is not carpentry. Once a life is touched, there is no rewind.

In synchronised, forward-thinking organisations, Quality by Design (QbD), integrated risk management, and human-centred design are not post-rationalisations. They are the nucleus of design excellence. They move us from simply meeting the bar to lifting the standard: for innovation, safety, and trust.

To design for trust is to begin with the end in mind, not merely a product, but a promise delivered under pressure, uncertainty, and scrutiny. It is a discipline, one that requires courage, foresight, and the humility to design with, not just for, those we serve.

In this chapter, we shift our perspective. We move from processes to purpose, from compliance to confidence. We explore the systems and mindsets that allow trust to be built into the blueprint, not inspected in after the fact.

Because in MedTech, trust is not a luxury. It is the most valuable asset we create and the most fragile.

Let us begin by making trust not just a byproduct of good design, but its very foundation.

8.1.1 Quality by Design – More Than a Methodology

Quality by Design is not simply a methodology, it is a philosophy. It is the radical belief that quality is not discovered late. It is created early. It is not derived from mere observation but rather arises through informed insight.

QbD challenges teams to think differently. To ask not what we can build, but what we must build for whom and why. It invites foundational questions: Who will use this device? In what context?

Under stress or urgency? With what training, in what setting, at what risk?

When embraced early, QbD transforms the development journey from reactive to proactive. Requirements become anchors of excellence, not constraints. Risk becomes a lens for clarity, not a checklist of fear. Human factors evolve from compliance tasks into acts of empathy and foresight.

The result is not just a compliant device. It is a trusted companion in the hands of users: intuitive, reliable, and ready when it matters most.

> *"Designing for compliance may avoid penalties.*
> *Designing for trust ensures purpose."*

8.1.2 Risk Management – The Guardian of Good Design

In a truly synchronised system, risk management is not an auxiliary process. It is the discipline of foresight. It is the guardian that walks alongside design, not to constrain it, but to protect its intent.

From the outset, every decision must be tested through the prism of risk. Not only technical risk, but clinical, operational, reputational, and human risk. To ignore one is to invite blind spots, and to engage all is to design with wisdom.

The best organisations do not simply document risk; they seek it, surface it, and strategies around it. They treat risk as a creative input, not a bureaucratic burden. They understand that robust design is forged not in the absence of risk, but in the presence of risk well understood.

> *"It is not the risk you see that will derail you.*
> *It is the one you chose not to look for."*

8.1.3 Human Factors – Engineering Empathy

Where risk guards the system, human factors give it soul. In MedTech, the end user is not a theoretical construct, they are a person navigating complexity, vulnerability, and urgency.

Human factors remind us that even the most brilliant technology will falter if it cannot be used safely, instinctively, and effectively. Design must accommodate the realities of trembling hands, time-starved clinicians, and distracted caregivers. Good design speaks clearly. Great design disappears into intuition.

Human factors are not about simplicity for its own sake. It is about clarity under pressure. Trust is established not solely through functionality, but also through the perception that a device comprehends and effectively supports its user.

When teams internalise this principle, they stop designing for approval. They start designing for real life.

8.1.4 The New Standard of Trust

Designing for trust is no longer a competitive advantage. It is the minimum requirement in a world where scrutiny is constant, expectations are rising, and the cost of failure is unacceptably high.

This chapter will explore how QbD, risk management, and human factors intersect not as academic frameworks, but as operational disciplines. Disciplines that allow us to design with clarity, deliver with integrity, and lead with confidence.

Because in MedTech, trust is not a final milestone. It is a continuous design outcome: earned, lost, and earned again.

And the organisations that embrace this truth do not merely meet expectations. They redefine them.

8.2 Quality by Design: Getting It Right the First Time

In the world of medical technology, designing it right the first time is not just a mantra. It is a moral imperative. Quality by Design is the conscious refusal to let chance determine safety. It is the discipline of foresight, embedded early, expressed consistently, and revisited continuously.

Designing for quality begins long before the first prototype is milled. It begins with conviction that a device should not only function but inspire confidence in those whose lives depend on it. QbD is not a toolset. It is a mindset. And it begins with the foundation.

8.2.1 Pouring Quality into the Foundation

Every enduring structure draws strength from what lies beneath. The same is true of great medical devices. Their integrity is not added at the end but poured into their foundation, long before a physical form exists.

QbD challenges us to design not for minimum viability, but for maximum reliability. It transforms quality from an outcome into an origin. Every early decision about materials, controls, interfaces, and even documentation is an opportunity to embed robustness and clarity where it matters most, at the root.

"You cannot inspect quality into a product. It must be built in."

This timeless principle, often echoed by pioneers like W. Edwards Deming, carries particular gravity in MedTech, where the price of failure is measured not in dollars, but in human impact.

8.2.2 Defining Quality Early – Setting Targets That Matter

A synchronised organisation does not guess what quality means. It defines it early, precisely, and courageously.

This begins with translating intent into design-critical targets. Take, for instance, an implantable device designed not just to perform, but to endure. The team sets a benchmark: 99.5 percent survival at five years. This target becomes a compass. It shapes material science, supply chain tolerances, test methodologies, and even packaging strategy.

Clarity is not restrictive. It is liberating. It aligns engineering ambition with regulatory realism and business purpose. It transforms conversations. Suddenly, trade-offs are not arguments, they are decisions made in the light of shared intent.

When quality targets are visible, measurable, and meaningful, they become the scaffolding for trust.

8.2.3 Design Inputs - Building Bridges, Not Barriers

QbD thrives on dialogue. In high-performing teams, design inputs are not isolated deliverables passed from silo to silo. They are the product of continuous negotiation between disciplines, priorities, and perspectives.

In this model, marketing no longer prescribes requirements from a distance. Engineering no longer optimises for performance without understanding clinical realities. Quality is no longer the final gate, rather it is present at the first gate.

Regulatory intelligence, serviceability, human factors, sustainability – all these voices shape the blueprint from the outset. Each discipline brings not constraints, but insight.

As I often remind teams embarking on design reviews:

*"These meetings are not performance evaluations.
They are rituals of trust, designed to reveal
assumptions and align around purpose."*

8.2.4 Design for X – Trust Into Every Layer

Design for X (DfX) is not a checklist. It is an operating system for quality.

Design for Manufacturability ensures that elegant concepts survive real-world production. Design for Reliability embeds resilience into every function. Design for Usability transforms complexity into clarity. When these lenses are applied early and often, they reveal failure modes before they ever reach a patient.

Imagine a team choosing between two component options: one sleek and unproven, the other simple and serviceable. DfX does not reject ambition. It demands evidence. It asks: will this choice still perform on the hundredth use, in the hands of a stressed clinician, in a remote setting, under uncertain power conditions?

To design for trust is to design for reality. It means making peace with complexity but not worshipping it.

8.2.5 A Living Example – Designing for Trust Upfront

Consider the journey of a MedTech company building a next-generation insulin pump. In a bold move, the team began their development with a "pre-mortem," a disciplined exercise in imaginative failure. What could go wrong? How might the user suffer?

They identified potential battery failure, user confusion, and even rare but catastrophic dose errors. These insights were not stored in a report. They were transformed into design action: multi-layered alerts, intuitive safeguards, and adaptive dosing algorithms.

By launch, the device did not merely pass regulatory scrutiny. It earned user loyalty. Why? Because the team had designed for failure, not just success, and had engineered trust into every layer.

8.2.6 Feedback Loops – When Learning Is Designed In

QbD is not only about getting it right the first time. It is about getting better every time.

A truly mature design system does not treat launch as the end. It treats it as the beginning of a new learning loop. Real-world use generates data. Field insights generate nuance. Complaints become clues. And these are not burdens, they are gifts.

Closing the loop means building pathways from post-market experience back into design. It means that every complaint reviewed, every deviation analysed, every risk reassessed becomes fuel for better requirements, stronger specifications, and wiser trade-offs.

Organisations that embrace this loop move faster, with fewer failures and greater resilience. Because their system is not just built for design, it is built for renewal.

8.2.7 Closing Thought

QbD is not a promise of perfection. It is a commitment to preparedness. It demands a mindset that values early clarity, rigorous empathy, and continuous learning. It does not guarantee a flawless path, but it ensures that even when things go wrong, the system knows how to respond.

As we now step into the domains of risk management and human factors, carry this forward: quality is not built in isolation. It is synchronised through systems, sustained by curiosity, and delivered by courageous teams.

In MedTech, the future belongs to those who design not just products, but trust. One requirement, one review, one release at a time.

8.3 Risk as a Design Driver: ISO 14971 in Action

Designing for trust means acknowledging that risk is not a roadblock, it is the road map. In fields where irreversible harm and the preservation of life are key considerations, risk management is a primary concern rather than a minor detail. It is the blueprint.

Risk management is sometimes viewed as a retrospective activity, carried out primarily as a procedural requirement before submission. In world-class organisations, this view has long been obsolete. Risk is no longer a form. It is a mindset. And ISO 14971 is not a checklist. It is a compass.

8.3.1 Risk as a Companion, Not an Afterthought

True design teams do not ask how to make a product work, they ask how it might fail. And they do it early, often, and without ego.

In the mature MedTech enterprise, risk is a companion that walks alongside the innovation process from the very first sketch. It asks uncomfortable questions. It sees around corners. It anticipates not only mechanical breakdowns, but also human missteps, unexpected contexts, and edge-case scenarios.

ISO 14971 guides this partnership through a deceptively simple structure: identify hazards, estimate risks, implement controls, and verify their effectiveness. When applied with intention, it creates not just safer products, but wiser ones. It ensures that innovation and caution do not sit on opposite sides of the table. They design together.

In this model, risk is not the enemy of progress. It is the architect of responsible progress.

8.3.2 The Living Risk File – From Concept to Completion

Risk management does not begin at design lock. It begins at concept. It lives through feasibility, development, and commercialisation. And it evolves post-launch, because risk itself is dynamic. It changes as usage scales, as user groups shift, and as unforeseen behaviours emerge in the real world.

The best organisations understand this. Their risk files are not dead archives. They are living systems. They are updated not only with verification test data but with insights from usability studies, complaints, service calls, and post-market surveillance. In doing so, the product continues to learn, even after it ships.

> *"The risks you ignore during development*
> *will find you during deployment."*

When risk perception is treated as fluid, not fixed, it becomes a catalyst for design agility. What seemed a minor hazard in controlled hospital settings might evolve into a major concern in a home-care environment. This is not a flaw in process. It is a feature of maturity.

8.3.3 Risk Controls – Designing Safety In

A cornerstone of ISO 14971 is its insistence on controlling risk by design, not just by warnings or procedural reliance.

The message is clear: do not ask the user to carry the burden of safety. Carry it for them through thoughtful, resilient design.

A robust medical device should make unsafe use unlikely, if not impossible. If a product requires the user to follow instructions to avoid harm, then it is not finished. It is only framed. True safety comes when the right way to use the device is also the only way it can be used.

The discipline here is engineering-led, not documentation-led. And the question is always the same: how can we make this harder to misuse, without making it harder to use?

8.3.4 Linking Risks to Design Outputs

The power of risk management is not in abstract matrices. It is in action. And action means linkage.

Every identified hazard should trace cleanly to a design output: a feature, a control, a specification, a test. This traceability ensures that risk management becomes an engine for clarity, not a parallel process but a fused one.

Take, for instance, the hazard of overheating in an electromechanical device. Risk thinking doesn't stop at identifying the hazard, it informs the solution: a sensor, a threshold, an automatic shutdown. All designed in, not written about. All validated, not theorised.

When risk is embedded in requirements, reflected in outputs, and validated in real use, it becomes more than managed. It becomes mastered.

8.3.5 The Price of Neglect – A Story of Two Infusion Pumps

History teaches. And it teaches through contrast.

One manufacturer of infusion pumps faced a known hazard: incorrect cassette loading could cause free flow, leading to dangerous overdoses. Their response? Add a prominent warning. Reinforce training. Trust the user.

A rival, facing the same risk, chose instead to trust the design. They reengineered the cassette so that incorrect loading was mechanically impossible. No instructions needed. No error possible.

The first company learned the cost of neglected design through patient harm, public scrutiny, and eroded trust. The second earned market share and regulatory praise, not through better marketing, but through better decisions.

This is not a cautionary tale. It is a call to action.

8.3.6 Closing Reflection

Risk is not a constraint. It is a responsibility. And in the hands of committed teams, it becomes a competitive advantage.

The most innovative organisations are not fearless. They are fear-aware. They embrace risk not to limit imagination but to shape it towards products that are not only new, but necessary. Not only safe, but trusted.

When ISO 14971 is embraced as a compass – dynamic, participatory, and practical – it no longer feels like governance. It feels like wisdom.

As we now move into the realm of human factors, let us remember it is not enough for devices to be safe in theory. They must be safe in practice, in real-world settings, with real-world users. And that means designing not only for risk, but for people.

8.4 Human Factors Engineering: Designing for People, Not Just Specs

8.4.1 Beyond Performance: Designing for Real Life

A device may pass every test, tick every box, and earn regulatory approval, yet still fail catastrophically when placed in a human hand. This is the unspoken irony of medical technology. Devices do not live in design files or static validation reports. They live in the unpredictable, emotionally charged world of real use, in hospitals at midnight, in homes after difficult diagnoses, in moving ambulances, in the hands of people navigating fear, fatigue, and urgency.

Human Factors Engineering (HFE) is not a compliance formality. It is the bridge between technical brilliance and human reality. It ensures that a product not only functions but also feels familiar, forgiving, and trustworthy under pressure. Because trust is not built in the specs. It is built in the moment someone uses the device for the first time and finds it guides rather than confuses, supports rather than intimidates.

This is not an act of charity. It is a design imperative.

8.4.2 Regulatory Push and Practical Wisdom

Regulators, led by the FDA and increasingly mirrored by global authorities, have made human factors validation a non-negotiable standard for critical devices. Summative usability testing, once considered a best practice, is now a regulatory floor.

But the most forward-looking organisations see beyond compliance. They view human factors not as a hurdle, but as a strategic differentiator. They move upstream, bringing formative testing into the earliest stages of design. Real users, real environments, real prototypes. Observing where people hesitate, misinterpret, or become frustrated becomes a lens through which smarter, more humane decisions are made.

It is a shift in mindset from blaming errors to uncovering insights. Every deviation isn't just a flaw to correct, but a cue to design more intelligently.

As I often say:

> *"Regulation is the minimum. Dignity,*
> *safety, and clarity are the goal."*

8.4.3 Designing for Trust through Simplicity and Clarity

Devices that are truly trusted do not merely meet functional requirements. Rather, they create an emotional contract with the user. That contract is written in simplicity.

An infusion pump that visually confirms settings in one glance. A ventilator alarm that instantly communicates what matters most, without ambiguity. A diagnostic app that leads users step-by-step, without confusion, especially under stress. These are not luxuries. They are the foundation of trust.

Design for humans means designing for reality: harsh operating theatres, noisy ICUs, dimly lit homes. It means assuming that the user may be tired, anxious, multitasking. And it means ensuring the product still works, guides, and protects.

The most successful devices are not just understandable, they are impossible to misunderstand.

8.4.4 Illustrative Scenario: The Home Dialysis Machine

Imagine a team developing a home-use dialysis system, empowering patients to take control of their care. Early usability testing revealed that users frequently connected two critical tubes incorrectly. It was a simple, subtle mistake with profound consequences.

Rather than respond with another warning label or longer instructions, the team redesigned the connectors entirely. Colour-coded, shaped to only fit one way, tactilely distinct. The error was designed out, not trained away.

Post-launch surveys revealed a powerful result. Users felt safer. They were more confident. The device had not only earned regulatory approval, it had also earned trust.

This is the essence of human factors: design that respects people, not by expecting perfection, but by anticipating imperfection with grace and precision.

8.4.5 Emotional Design – Trust at a Glance

Before a device is even switched on, it begins telling a story.

Is it sturdy or fragile? Cluttered or clean? Confusing or calming? These visual and tactile cues shape emotional response before any functionality is proven. The user has not read the manual yet, but they have already decided whether to trust it.

This is not cosmetic. It is critical.

The best MedTech products are designed as if someone's life depends on it, because it often does. The design of their form conveys a clear sense of purpose. Their layout reassures. Their colours and materials speak softly of safety, calm, and confidence.

> *"Design as if the person using your*
> *device were your own loved one, because*
> *someone's loved one will use it."*

Designing for emotion is not soft. It is strategic. A device that instils confidence is more likely to be used correctly, consistently, and comfortably. It becomes not only usable but indispensable.

8.4.6 Closing Reflection

Human Factors is not a layer added at the end. It is the soul of quality, embedded at the beginning. It invites us to see through the eyes of users, to walk in their shoes, and to design with empathy, not just engineering prowess.

The future of MedTech will not be defined by speed alone or by specifications achieved. It will be shaped by how well we listen, how wisely we anticipate, and how deeply we design for the people we serve.

Let us build not only devices that work, but experiences that heal. In doing so, we will not just meet the standard. We will redefine it.

8.5 Bringing It Together: QbD, Risk, and Human Factors

8.5.1 The Power of Connection

When we examine the world's most trusted medical technologies, we rarely find brilliance born from a single stroke of genius. Instead, we encounter something far more disciplined: a convergence of capabilities, patiently layered and consciously connected. The secret behind these devices is not luck, nor even rigorous science alone. It is the intentional fusion of three essential disciplines: Quality by Design, Risk Management, and Human Factors Engineering.

These are not parallel processes competing for priority. They are symbiotic principles. Each one, when deeply embedded, reinforces the others. QbD anticipates excellence. Risk Management safeguards it. Human Factors humanises it.

Trust, after all, is not the sum of separate actions. It is the outcome of coherent intent.

8.5.2 Designing with Risk and Humans in Mind

A colour-coded connector on an infusion pump. At first glance, it seems a small gesture, perhaps a nod to usability. But look again. That connector prevents misconnections, reduces harm, and communicates intent without words. It is a risk control. It is a QbD choice. It is a Human Factors insight.

This is where the disciplines converge, where the so-called "soft" elements of empathy meet the "hard" imperatives of safety. And it all begins at the sketchpad. Not in the test lab. Not in the regulatory file.

By designing with both risk and the human experience in mind, teams move from reaction to anticipation. They build defences into the product before vulnerabilities can take root. They do not wait for failure to reveal design flaws. They listen early, learn constantly, and build trust intentionally.

8.5.3 Breaking Silos by Necessity

This convergence cannot occur in organisational isolation. It demands the dissolution of silos: technical, procedural, and cultural. It requires quality leaders to speak with engineers before concepts are frozen. It asks Human Factors experts to inform design architecture. It invites clinicians, regulatory professionals, and even marketing voices into the formative stages.

This shift does not happen by policy. It happens by ritual.

Cross-functional design reviews where usability and risk sit side by side. Scenario-based simulations with "human-in-the-loop" stress testing. Risk-based brainstorming workshops that include not just "what could go wrong?" but "how might a user misunderstand?" Shared ownership of early failures and shared celebration of early insights. These practices are not overhead, they are the scaffolding of trust.

"You cannot synchronise quality, safety, and trust if the people responsible for them only meet after the design is done."

8.5.4 Sailing Through Verification, Validation, and Approval

When this harmony is achieved, the later stages become a confirmation, not a correction.

Verification no longer uncovers late-stage surprises, it affirms decisions made wisely and early. Validation becomes less about stress-testing assumptions and more about demonstrating that real-world needs were built in from the beginning. Even regulators, often seen as barriers, become partners. They recognise the rigour. They feel the respect.

A submission built on QbD, comprehensive risk thinking, and deep human insight does not merely answer the required questions, it pre-empts them. It creates an unmistakable signal: this is a product designed not just to meet expectations, but to earn confidence.

8.5.5 False Economy – A Reformer's Warning

And yet, the temptation persists. The shortcut. The quick path to prototype. The assumption that a warning label can substitute for good design. That a training session can fix what the interface never made clear.

It is a seduction born of speed and scarcity. But it is a dangerous illusion.

Because the cost of not designing for trust is rarely paid in the development phase. It arrives later, in safety notices, in redesigns, in regulatory scrutiny, and in the quiet erosion of user confidence. And once trust is lost, it is rarely regained.

The real economy is found in getting it right before it goes wrong.

Fixing the blueprint costs less than fixing the recall. And more than that, it preserves the one currency we cannot afford to lose: belief. Belief from the clinician who picks up the product in the heat of crisis. Belief from the patient who places their future in its hands. Belief from the regulator who must decide not just whether it complies, but whether it cares.

8.5.6 Final Reflection: Designing Belief into the Blueprint

Trust is not a feature. It is not something layered on top or added in a final pass. It is what happens when design integrity, safety thinking, and empathy are synchronised from the beginning.

The synergy of QbD, Risk Management, and Human Factors does more than produce great devices. It redefines what "good" means in MedTech. It builds trust not by chance or charm, but by structure.

And that trust becomes the foundation not only of safer products but of healthier, stronger organisations.

The next chapter of excellence will not be written by the fastest or the flashiest. It will be led by those who understand that the only way to scale innovation in healthcare is to make trust non-negotiable by designing it into every detail, every decision, every device.

Let us commit to that standard. And let us build not just compliance, but conviction: one product, one patient, one purpose at a time.

8.6 Embedding Trust into the Product DNA

8.6.1 The Blueprint Beneath the Surface

When a medical device earns admiration for its precision, safety, and grace, what is seen is only the surface, the final manifestation of choices made long before the first prototype emerges. Trust, in truth, begins quietly. It is seeded in the margins of whiteboards, in the early meetings where ideas collide and priorities are shaped, in the tension between bold innovation and careful restraint.

The soul of a trustworthy product is formed in those moments. It is forged in the debates over trade-offs, the thoughtful articulation of what "quality" means, and the courage to ask, "What could go wrong and how will we prevent it?"

> *"The lion's share of product quality is determined in the design room, not the inspection room."*

What follows is merely validation. The real work – of trust, of safety, of excellence – happens beneath the surface.

8.6.2 Trust Built at the Blueprint Stage

Devices designed through the intentional integration of Quality by Design, robust Risk Management, and deep Human Factors insight do more than pass tests. They resonate. They intuitively align with users' needs, glide through verification, and earn regulatory approval not through clever arguments, but through evident diligence.

Post-market data reflects this care. Fewer adverse events. Higher user satisfaction. Stronger brand equity. Teams spend less time firefighting and more time evolving, refining, improving. When trust is built into the foundation, it becomes a self-sustaining cycle. Confidence fuels adoption, adoption drives feedback, and feedback strengthens the next iteration.

This is the quiet compounding power of thoughtful design.

8.6.3 The Return on Thoughtful Design

There is a myth in MedTech that time spent in early design delays success. But history and experience prove the opposite. Every hour invested in defining quality, interrogating risk, and designing for human reality pays back in avoided rework, regulatory clarity, and reduced market risk.

Design diligence is not an overhead. It is strategic acceleration.

The leaders of tomorrow are not those who ship first, but those who ship with integrity. They are the ones who internalise that trust is not something you retrofit. It is something you embed, by choice, from the beginning.

"If my post-mortem reveals a device failure, I
refuse to be just another complaint in a database.
Build it right, because lives are not statistics."

8.6.4 Setting the Stage for Execution

This chapter has made one truth clear: a trustworthy device is not born in the testing lab or the boardroom. It is born in the earliest design decisions, in those quiet, foundational moments that rarely make headlines but shape everything that follows.

Because in MedTech, trust is not something you test.

"Trust is something you build,
before you test anything."

And when it is built right, the product becomes more than a device. It becomes a promise. A standard. A legacy.

This is not just a call to reflect. It is a call to design with intent. Trust is not tested in, rather it is designed in from the start. Ask yourself and your teams bold, practical questions:

- Are your product development processes integrating QbD, risk, and Human Factors from concept to launch or treating them as checkboxes?
- How well do your teams understand the real-world context of the user, not just the technical specifications?
- Are you using risk management (ISO 14971) to drive design decisions, or just to justify them after the fact?
- Where could earlier integration of Human Factors have prevented post-market issues or usability complaints?

In the next chapter, we enter the digital frontier. We explore how intelligent technologies are redefining the MedTech lifecycle and how synchronisation remains the critical enabler. From Agile methodologies to AI integration, we look at what it takes to harmonise innovation with compliance in a world that moves at digital speed.

CHAPTER 9

DIGITAL LIFECYCLES: ORCHESTRATING INNOVATION IN SOFTWARE, AI, & SMART DEVICES

There was a time when a medical device meant something mechanical: tangible, engineered, precise. Innovation was measured in materials and mechanics. But that era is evolving. Today, the most transformative components in MedTech are often invisible: software code, learning algorithms, and connected data systems. The frontier of innovation has shifted from the physical to the digital and with it, so must our thinking.

We are no longer building static products. We are shaping living, adaptive ecosystems, intelligent systems that learn from real-world data, update in real-time, and integrate seamlessly across care environments. The traditional lifecycle of design, build, verify, launch is giving way to something far more fluid, continuous, and collaborative.

This shift brings immense opportunity, but also profound responsibility. As AI and software redefine diagnostics, monitoring, and intervention, they also introduce new layers of complexity. These systems don't just execute, they evolve. And when systems learn, we must ensure they remain safe, ethical, transparent, and accountable.

This chapter explores how MedTech organisations can thrive in this new paradigm, where quality becomes a continuous rhythm and risk management transforms from a static documentation exercise into a living, responsive discipline.

You will discover:

- How to govern AI and data with the same rigour once reserved for hardware
- Why post-market data is not just surveillance, but the fuel for evolution
- How to embed cybersecurity, privacy, and ethics directly into your QMS
- And how to build systems that don't just function but adapt, connect, and improve with every interaction

Because in MedTech, the intelligence of your technology is only as powerful as the integrity of the system that supports it.

At the heart of this transformation is synchronisation. Agile development, design control, regulatory oversight, and real-world feedback can no longer operate in silos. They must converge into a unified operating rhythm, one that balances speed with safety, change with control, and innovation with trust.

> *"In digital MedTech, excellence isn't what you launch. It is how consistently you evolve with clarity, control, and care."*

Welcome to the digital lifecycle, where updates are no longer afterthoughts, but acts of continuous responsibility. Where agility is governed, data is purposeful, and innovation is always in sync with compliance.

This is the future of MedTech, and it is already in motion.

9.1 Where MedTech Meets Silicon Valley: The New Lifecycle Paradigm

9.1.1 The Future Now Arrives with a Software Update

Once, medical devices followed a predictable lifecycle: design, validate, approve, launch, then run until obsolete. Their performance was static, their functionality fixed. Today, we have crossed a threshold. Devices learn, update, and evolve, sometimes overnight.

We have entered the age of software-defined MedTech, where a patch isn't just a technical update and can be a clinical improvement. Surgical robots refine their precision. Diagnostic systems adapt from new data. Pumps enhance safety protocols via real-time cloud feedback. The "product" is no longer frozen at launch, it is alive.

This is more than a technical shift; it is a philosophical one. Devices are no longer seen as finished artefacts but as evolving capabilities. As Marc Andreessen once noted, "Software is eating the world." In MedTech, it is no longer nibbling, it is reprogramming the very definition of care.

For regulators, quality professionals, and innovators alike, this requires a new orientation, where the question is no longer "is it ready?" but "can it stay ready?"

9.1.2 From Static Pipelines to Dynamic Ecosystems

The traditional lifecycle: Design, Develop, Validate, Launch, Monitor was built for hardware, with clearly defined gates and stages. But today's MedTech products are no longer stand alone. They are ecosystems: combinations of apps, algorithms, cloud services, and real-time patient interactions.

Each component influences the others. A software update can ripple through cybersecurity protocols, clinical workflows, and user interfaces. Change is constant. Coordination is no longer a best practice, rather it is a survival strategy.

Lifecycle management must become cyclical, responsive, and co-owned. Product teams must abandon the myth that "done" is a moment. In digital MedTech, done is a moving target, and that is a good thing if systems are designed to evolve safely.

Regulatory and quality models must keep pace. Instead of resisting change, they must enable iteration while safeguarding trust.

"The future belongs to those who can
evolve without eroding trust."

9.1.3 Synchronising Speed and Trust: Questions for a Digital Future

Speed alone is not a virtue. A glitch in a consumer app is an inconvenience but in healthcare, it can be fatal. The challenge isn't whether to move fast, it is how to move fast responsibly.

This is where synchronisation becomes the critical differentiator. Regulatory foresight, agile development, quality systems, and cybersecurity can no longer be sequential. They must operate in concert. Traceability must be automated. Risk thresholds must be pre-defined. Decision logic must be auditable.

This is not about slowing down. It is about rethinking governance for motion.

Forward-leaning organisations are embedding regulatory voices into every sprint and adopting modular architectures and Software Bills of Materials (SBOMs) to manage dynamic cybersecurity.

> *"Speed without structure is chaos. In*
> *healthcare, chaos costs lives."*

9.1.4 The Cultural Shift

Perhaps the greatest transformation is cultural. In the old paradigm, innovation and compliance were perceived as opposites. Today, visionary organisations are building cultures of digital trust where speed and safety, iteration and integrity coexist.

Risk is managed in real-time. Quality is integrated, not inspected. Innovation is not a departure from discipline, it is its evolution.

This demands unlearning. Legacy models must be dismantled. Mindsets must shift from "launch readiness" to constant readiness. Teams must design for movement, not maintenance.

> *"Excellence is no longer a moment, rather it is*
> *a rhythm. A commitment to stay ready, update*
> *responsibly, and earn trust continuously."*

9.2 Why Software and AI Demand a Different Playbook

9.2.1 The Shape of Complexity We Can't See

The complexity of modern MedTech is no longer confined to metal, plastic, or firmware. It now lives in code, algorithms, cloud integrations, and intelligent networks. This complexity is invisible, interconnected, and constantly evolving, and that makes it dangerous when poorly understood.

Today's devices are no longer standalone tools, they are intelligent nodes within a dynamic, interconnected digital ecosystem. Apps, EHRs, wearables, real-time patient feedback – everything interacts. And every interaction is a potential point of failure or learning.

Traditional quality systems were designed for static hardware. They assumed rare, controlled change. But now, we must govern continuous change, and that requires more than traceability, it demands synchronisation across design, risk, cybersecurity, and real-world performance.

9.2.2 Managing the Digital Unknown – Risk in a Living System

In this new paradigm, change is constant and not always visible.

- Software updates
- AI model retraining
- External dependencies (e.g. EHR integrations)
- Cloud API updates
- Shifts in patient data

All these can silently introduce risk. A perfectly functioning device can degrade, not due to a defect, but because the ecosystem around it has changed. Welcome to model drift, cyber drift, and ecosystem misalignment.

Risk management must move from periodic reviews to continuous readiness. That means:

- Dynamic threat modelling
- Post-deployment cybersecurity monitoring
- Real-world data performance tracking
- Interoperability testing embedded into development
- Third-party mapping and Software Bills of Materials (SBOMs)

> *"Resilience is the new reliability. And in digital MedTech, resilience means designing systems that remain safe through constant change."*

Traditional methods, retrospective FMEAs or annual reviews, cannot manage this kind of complexity. A living system demands a living risk process.

9.2.3 New Compliance Expectations

Regulatory frameworks are catching up and, in many ways, redefining the rules.

- IEC 62304 (software lifecycle)
- ISO/IEC 27001 (cybersecurity)
- ISO 14971 (risk)
- FDA's AI/ML framework

- EU AI Act
- ISO 42001 (AI management systems)

Compliance is no longer just about documentation, it is about showing how systems evolve safely. Static validation is not enough. Organisations must embed traceability into development, govern updates proactively, and anticipate audit trails across machine learning models.

Too many legacy organisations try to bolt AI onto old systems. But that is like installing autopilot into a horse cart.

What is needed is a complete re-architecture of quality and regulatory processes for software-first, AI-native products:

- Validation for continuous updates
- AI explainability and performance guardrails
- Data governance (lineage, security, anonymisation)
- Model drift detection and retraining logic
- Embedded cybersecurity-by-design

> *"In a world of learning systems, trust isn't granted at launch. It is earned with every update, every input, and every outcome."*

9.2.4 The New Role of the QMS: From Gatekeeper to Guide

This shift redefines the role of the QMS. No longer a final checkpoint, it becomes a navigation system constantly guiding, interpreting, and adjusting based on real-world signals.

The synchronised QMS of the future isn't reactive. It is predictive. It doesn't just document change, it orchestrates it.

When quality, regulatory, data, and design teams work in sync, not in sequence, MedTech can move at the speed of software without sacrificing the safety of medicine.

9.3 Agile Meets Regulation: Harmonising Speed with Control

9.3.1 The Myth of Misalignment

One of the most persistent misconceptions in MedTech is that Agile and regulation are incompatible, that speed and structure are natural enemies. In truth, they are co-requisites in a world where software evolves daily and patient safety cannot wait.

Agile development brings adaptability, user-centricity, and rapid iteration. Regulation brings safety, reliability, and traceability. The challenge isn't choosing one over the other, it is designing systems where both can thrive.

> *"Agility without accountability is chaos.*
> *Accountability without agility is stagnation.*
> *Synchronisation is the bridge."*

Synchronised organisations are no longer asking, "Can we use Agile?" They are asking, "How do we embed Agile into a system that ensures continuous compliance and trust?"

9.3.2 Agile Under Design Control: Compliance, Sprint by Sprint

When implemented thoughtfully, Agile doesn't replace design control, it distributes it.

Each sprint becomes a micro-lifecycle: defining requirements, verifying functionality, reassessing risk, and delivering usable output. Compliance isn't deferred to the end, it is embedded into the flow.

Frameworks like AAMI TIR45 have shown regulators how Agile can support, not undermine, quality and traceability. Key enablers include:

- Modular documentation (user stories, sprint-based Verification & Validation (V&V), live risk registers)
- Sprint-level design reviews
- Risk updates as living artefacts
- Traceable links between inputs, outputs, and decisions

This turns Agile from a speed enabler into a compliance multiplier.

"Design control isn't diluted by Agile;
it is made continuous."

9.3.3 Test Automation – Agile's Quality Backbone

In traditional development, testing is a milestone. In Agile, it is a heartbeat.

Test automation enables rapid iteration without losing control. With all change, automated regression tests, performance checks, and verification scripts provide instant feedback and documented proof of functionality.

In a synchronised Agile-QMS, these automated tests serve two masters:

- They confirm product readiness
- They support compliance through traceability, repeatability, and auditability

This redefines "done." Readiness is no longer just about completing a feature, it is about delivering validated, trustworthy functionality.

9.3.4 The New Discipline

Adopting Agile in MedTech isn't about becoming less rigorous, it is about becoming rigorous in motion.

Documentation shifts from static plans to real-time evidence. Risk evolves with the product. Validation becomes continuous. And transparency is no longer a burden, rather it is a design principle.

This is the new discipline: Agile synchronised with quality and regulatory functions, forming a single, adaptive engine of innovation.

> *"In regulated MedTech, agility must never*
> *come at the cost of control. But done right,*
> *it becomes the fastest path to trust."*

9.4 Continuous Deployment vs. Compliance: Updating Without Risk

9.4.1 The New Frontier of Deployment

In the past, MedTech updates were rare, disruptive, and often met with regulatory friction.

But today, products evolve in real time. Software patches, algorithm refinements, and safety enhancements can and must happen continuously.

This new normal requires more than agile engineering. It demands regulatory-ready deployment systems that balance speed with vigilance. The question is no longer "Can we update quickly?" It is "Can we update responsibly and in harmony with compliance?"

Leaders in this space don't treat post-market updates as exceptions. They design for them, embedding readiness into architecture, governance, and QMS from day one.

9.4.2 Managing Change in a Digital World – Risk-Based Discipline

Not all updates are created equal. A UI enhancement differs dramatically from an AI model change impacting clinical decision-making. That is why a risk-tiered change control framework is essential.

Smart MedTech teams:

- Pre-classify update types (clinical, cybersecurity, algorithmic, UX, etc.)
- Define thresholds for when regulatory action is triggered
- Map each change to pre-agreed protocols: internal review, documentation, or submission

This structure is especially critical in over-the-air (OTA) environments, where software can change on thousands of devices simultaneously. It is not just about pushing the update, it is about orchestrating the system around it: rollback plans, audit trails, version control, and secure deployment.

"Real-time iteration demands real-time governance."

A synchronised QMS doesn't slow you down, it creates predictable agility.

9.4.3 Transparency – Building User Trust

Even the most elegant update loses impact if users don't understand it or, worse, don't trust it.

Change without clarity invites suspicion. But clearly communicated change reinforces trust. Each release should include:

- Accessible release notes tailored to users (clinical, technical, patient)
- Impact summaries (especially for safety or performance)
- Training nudges, where needed, to ensure updates translate to real-world benefit

In a world where software is the product, transparency isn't a courtesy, it is a core design feature.

9.4.4 Illustrative Scenario – Trust at Speed

Consider a connected glucose sensor. Within weeks of launch, field data revealed algorithm drift in specific humidity conditions.

Because the team had pre-designed for continuous deployment, they validated a patch, deployed it across all affected units in fourteen days, and issued a transparent user communication, all within a regulatory-approved change framework.

There was no crisis. No delay. No erosion of trust.

This is synchronisation in action. Agility with foresight. Updates with discipline.

9.4.5 Conclusion – Continuous Change as Competitive Advantage

Continuous deployment and regulatory control are not in conflict; they are complementary mechanisms in a unified system, each essential to sustainable innovation.

Organisations that build change-ready systems gain more than speed, and they earn enduring trust by transforming updates from disruptions into strategic moments of continuous improvement.

> *"In the digital lifecycle, excellence isn't
> a milestone. It is a rhythm: one update,
> one insight, one decision at a time."*

9.5 AI and Data Intelligence: What MedTech Must Consider

9.5.1 The Pulse of Modern MedTech

In today's MedTech landscape, software is the foundation, but AI is the nervous system. Devices don't just executing, they sense, learn, and evolve. Algorithms guide diagnosis, tailor therapies, and influence clinical outcomes in real time.

Yet with this intelligence comes uncertainty.

AI doesn't just change how products function, it transforms how they learn. And learning systems, by definition, are never finished. Traditional design controls, static validation, and linear updates simply can't keep up.

> *"Innovation without governance is
> fragility disguised as progress."*

The future of MedTech demands that AI be treated not as an add-on, but as core architecture, governed with the same precision as any critical system.

9.5.2 Trust Beyond the Algorithm

In MedTech, trust is the product.

For AI to be credible, it must be understandable, traceable, and explainable. Developers must not only know what the algorithm does, they must prove how, when, and why it changes.

This means moving from AI enthusiasm to AI stewardship.

Synchronised organisations embed:

- Diverse, real-world training datasets to reduce bias
- Explainability by design, not as an afterthought
- Model drift detection and retraining protocols
- Post-market surveillance tailored for AI performance
- Clear traceability from training data to decisions

Emerging standards like the EU AI Act, FDA's GMLP, ISO 42001, and IEC 62304 are setting new expectations, not just for safety, but for algorithmic responsibility.

"In learning systems, explainability is safety."

9.5.3 Data Governance – The Foundation of Digital Trust

No AI system is stronger than its data. And in healthcare, data is never just technical, it is personal, sensitive, and mission-critical.

High-performing organisations don't treat data governance as a compliance chore. They make it a design pillar.

This includes:

- End-to-end data lineage tracking
- Anonymisation and encryption by default
- Bias detection embedded in hazard analysis
- Governance that not only meets HIPAA and GDPR but also anticipates what comes next

When data is handled with integrity, it becomes fuel, powering safe innovation, faster feedback, and deeper insight. When mishandled, it becomes liability.

9.5.4 Designing for the Ecosystem, Not Just the Device

AI-powered MedTech doesn't live in isolation. It syncs with EHRs, mobile apps, cloud platforms, wearables, and more. A product is no longer just a product, it is a connected node in a living digital ecosystem.

This changes everything.

From day one, design must account for:

- Interoperability: using secure, standardised protocols (e.g., HL7 FHIR, DICOM)
- Modularity: enabling independent updates without destabilising the system
- Resilience: anticipating real-world variability in data, connectivity, and environments

And governance must keep pace, synchronising quality, regulatory, cybersecurity, and engineering teams across the evolving ecosystem.

In connected health, you don't just build the product, you build the relationships around it.

9.5.5 Final Reflection – The Ethical Code Beneath the Source Code

Digital excellence is not measured by performance alone, but by principles.

The most advanced MedTech solutions won't be those with the flashiest AI, but those that earn lasting trust through transparency, stewardship, and human alignment.

Synchronisation ensures AI doesn't just work but works ethically, designed with foresight, governed with care, and continuously improved with intention.

> *"In the age of intelligent machines,*
> *integrity is your sharpest edge."*

9.6 Bridging Cultures: When IT and MedTech Truly Collaborate

9.6.1 Bridging Two Mindsets for a Unified Future

The digital transformation of MedTech isn't just a technical challenge, it is a cultural one.

Traditionally, quality, regulatory, clinical, and engineering teams operated within structured hierarchies, guided by meticulous process discipline. Meanwhile, software and IT teams embraced Agile, cloud-native thinking, and rapid iteration. Both cultures were strong but rarely synchronised.

Now, the walls must come down.

Modern MedTech demands not just coexistence, but co-creation. It requires a shared language between those who ensure compliance and those who drive code. Synchronisation isn't a soft skill, rather it is a strategic necessity. Without it, innovation risks stalling in translation. With it, we unlock exponential capability.

> *"Integration isn't about compromise. It*
> *is about building a culture that is fluent*
> *in both precision and pace."*

9.6.2 The New Team Equation – Cultural Fusion in Action

When IT and MedTech speak the same language, product development transforms.

Gone are the days of regulatory being brought in at the eleventh hour or software teams operating outside of design control. Leading organisations are embedding regulatory foresight, risk analysis, cybersecurity, and human factors into Agile sprints, not as an overlay, but as native elements of the workflow.

This creates a new kind of team structure:

- Shared sprint rituals where QA and regulatory sit alongside dev and test

- Joint ownership of risks and requirements, not handoffs
- Continuous training loops where each discipline educates the other

This is not just collaboration. It is cultural fusion.

This alignment allows organisations to build systems, not just products – systems where trust, performance, and safety evolve together.

9.6.3 Real-World Integration – What 'Good' Looks Like

The most successful synchronised organisations aren't guessing, they are planning for integration from day one.

Examples include:

- Software teams using modular architectures to align development with regulatory submission blocks
- Regulatory functions adopting DevOps dashboards to track evidence generation in real time
- Quality teams shifting from gatekeepers to enablement partners, ensuring Agile doesn't mean ad hoc

At their core, these companies realise one thing: compliance is not a delay, it is a design parameter.

When synchronisation becomes operationalised, organisations move faster, with fewer surprises and higher trust, both internally and externally.

9.6.4 Closing Insight – Building More Than Products

Synchronisation isn't just about improving collaboration. It is about redefining what we build and how we build it.

In the digital age, a product isn't a finished object, it is an evolving service, a stream of updates, and a promise of safety. That promise is only kept when systems, people, and culture operate in harmony.

"Great MedTech isn't built in silos. It is
orchestrated by multidisciplinary minds
aligned around shared purpose."

The organisations that lead the next era will be those that transcend departmental boundaries, where IT, engineering, quality, and regulation are not departments, but a single nervous system.

When that happens, innovation scales. Risk becomes foresight. And every update becomes an opportunity to earn trust again.

9.7 Synchronising Velocity with Vigilance

9.7.1 A New Horizon for MedTech

We have entered a new era in MedTech, one where agility and accountability no longer exist in opposition but as co-requisites for leadership.

The traditional dichotomy, between speed and safety, between innovation and compliance, has fractured. In its place, a new standard is emerging: one where synchronisation is not a compromise but a competitive advantage. Quality, regulatory, and engineering departments must no longer chase each other in sequence. They must move together, as one.

"Synchronisation isn't the slowing down of innovation.
It is how innovation learns to move with integrity."

The companies shaping the future of healthcare will not be those who simply move fastest, but those who move wisely, responsibly, and in rhythm with the systems they serve.

9.7.2 Harnessing Digital Power Without Losing Our Soul

In a world of machine learning models, OTA updates, and real-time data feedback, it is easy to confuse technical velocity with progress. But in healthcare, speed without soul is risk. Every line of code, every algorithm update, every deployment must be underpinned by trust, not just technical confidence, but human trust.

Synchronised organisations understand this. They don't treat compliance as a box to tick, but as a signal of care. They embed ethical guardrails, privacy principles, and clinical rigor directly into their architecture. And they embrace agility, not to cut corners, but to close loops faster.

> *"Fast without control is chaos. Control without speed is stagnation. True excellence is momentum with mastery."*

9.7.3 A Living System of Learning

In the past, learning in MedTech was episodic, triggered by failures, recalls, or audits. Today, learning must be continuous. Every sprint, every signal, every update is a chance to get better, not just in the product, but in the system that supports it.

This is the true promise of a synchronised digital lifecycle:

- Post-market data doesn't just feed surveillance, it fuels design.

- Agile sprints don't dilute compliance, they strengthen it.
- Risk management isn't a checkpoint, it is a live conversation.

Quality is not the last stop before launch, it is the ongoing rhythm of responsible innovation. And regulation is not a limit, rather it is a lens for thinking more clearly about impact.

9.7.4 The Journey Continues

This chapter is not a conclusion, it is a threshold. What lies ahead is not simply more process, but more purpose.

Digital systems may power the future of MedTech, but it is human alignment that determines whether that future arrives seamlessly or splinters under the weight of fragmentation.

We have seen the blueprint: synchronising digital lifecycles, quality systems, and compliance frameworks demands more than technical redesign. It calls for cultural reinvention. Because even the most intelligent QMS or AI-enabled platform cannot realise its full potential if teams remain siloed, speak in different operational dialects, or cling to legacy structures that reward isolation over integration.

The true transformation begins when people – engineers, designers, regulatory leaders, clinicians, software architects, and quality professionals – begin operating as one synchronised ecosystem. When feedback becomes a welcome input, not a defensive trigger. When compliance is not someone else's job, but everyone's shared accountability. And when collaboration isn't an event, it is embedded in the culture.

"Innovation doesn't thrive in isolation. It thrives
in systems designed to evolve, together."

We now step into a new standard where speed is synchronised with safety, agility is aligned with assurance, and every update, from codebase to culture, earns the right to move MedTech forward.

This is not just a call to reflect. It is a call to rewire. The digital lifecycle requires a new rhythm, one that synchronises speed, safety, and system thinking. Ask yourself and your teams bold, practical questions:

- Are your current development processes designed for continuous iteration, or constrained by legacy waterfall thinking?
- Is there a clear governance model for validating and controlling changes in digital products without slowing innovation?
- Where do cultural divides between MedTech, IT, and software teams still create tension or risk?
- What steps are you taking to embed ethics, bias awareness, and data integrity into your AI or digital development practices?

As we turn the page, we shift from system design to human dynamics. In Chapter 10, we explore how high-performing MedTech organisations turn cross-functional collaboration from a buzzword into a disciplined, synchronised reality, resolving friction, aligning mindsets, and embedding collective intelligence into the DNA of every team and every decision.

Because in truly synchronised organisations, innovation isn't just executed, it is orchestrated.

SYNCHRONISING MINDS: UNITING CROSS-FUNCTIONAL TEAMS FOR MEDTECH MOMENTUM

Innovation in MedTech doesn't live within silos, it sparks at the intersections. It is born where disciplines collide, where engineers think alongside clinicians, and where quality, regulatory, and commercial minds challenge and champion each other. These intersections are not optional. They are where momentum begins.

Yet in many organisations, brilliance still operates in parallel rather than in partnership. Engineering crafts a world-class prototype, only for Regulatory to raise late-stage red flags. Clinical teams design trials in isolation, and Manufacturing scrambles during scale-up. Quality arrives just before launch, instead of shaping decisions from day one. The result? Rework, delays, diluted insights, and worst of all: unmet patient needs.

This chapter is about ending that pattern.

Because the best frameworks, systems, and tools mean little if the humans within them aren't aligned. True synchronisation isn't just process-deep, it is people-deep. It is not integration between software, rather it is integration between mindsets.

As Niels Pflaeging put it:

> *"Trust is the foundation of teamwork... but it*
> *is not the starting point. Trust is built through*
> *the act of working together, for each other."*

In the highest-performing teams, trust isn't a buzzword, it is the operating system. What drives results isn't hierarchy, but shared ownership. Not just coordinated calendars but connected convictions.

Multidisciplinary synchronisation isn't about meetings. It is about movement. It is about creating a culture where Quality, R&D, Regulatory, Clinical, Manufacturing, and Commercial don't just speak; they think together, plan together, and solve together. Where differences in perspective aren't just tolerated, they are treasured. And from that cognitive diversity emerges what the MedTech world needs most: innovation that is safe, fast, and trusted.

In the pages ahead, we will explore how to build those teams, ones that don't just perform tasks together, but create value together. Teams where synchronisation is not a buzzword, but a daily discipline.

Because when people align, innovation accelerates.
And when innovation is human-powered, trust becomes unstoppable.

Let us begin.

10.1 The Case for Synchronisation: Unlocking Collective Intelligence

Innovation in MedTech no longer unfolds in clean handoffs or isolated lanes. It happens in synchrony where engineering, clinical, regulatory, quality, and commercial minds converge in real time, not in sequence.

The complexity of modern health technologies, from AI-enabled diagnostics to connected therapeutics, demands more than functional excellence. It demands orchestration, a deliberate shift from individual brilliance to integrated intelligence. Because in today's environment, a perfectly engineered device that fails regulatory scrutiny, or a clinically validated system that can't be manufactured at scale, is not a breakthrough. It is a breakdown.

> **"A synchronised system without synchronised people is just an expensive illusion."**

10.1.1 Why Synchronisation Now?

Legacy product development models were built for a different era, when physical products moved predictably through silos: engineering designed, quality checked, regulatory approved, manufacturing scaled. That waterfall worked when timelines were longer and products were simpler.

But today's MedTech is a living system. Features evolve post-launch. Data flows bi-directionally. Compliance expectations are dynamic. The idea of "done" has been replaced with "always adapting."

Synchronisation isn't just a best practice, it is the only way forward.

Consider a connected drug delivery device:

- Engineering must design not only for performance but usability.

- Regulatory must anticipate evolving AI oversight requirements.
- Clinical must validate safety under real-world variability.
- Cybersecurity must be baked into design, not bolted on later.

If these perspectives don't shape the product together, from the start, the cost is not just rework. It is risk to patients, to market access, and to trust.

10.1.2 From Cross-Functional to Cross-Intelligent

It is easy to claim "cross-functionality." It is harder to live it.

Most organisations assemble teams with the right titles but still operate in functional silos. True synchronisation isn't about who is invited to the meeting. It is about who shapes the thinking.

- Human Factors joins early sketch reviews, not just summative testing.
- Quality co-authors verification plans, instead of just auditing them after execution.
- Regulatory advises on risk logic during feature iteration, not once the file is frozen.
- Clinical guides use scenarios, instead of just validating endpoints.

This is not a relay race. It is formation flying where each discipline adjusts in real time, with shared purpose and mutual awareness.

"Integration doesn't mean proximity. It means shared responsibility for outcomes."

10.1.3 Designing Teams for Synchrony

High-performing collaboration doesn't emerge by accident. It is designed.

Leaders must model and reward behaviours that reinforce alignment:

- KPIs that reflect system outcomes, not siloed wins.
- Decision forums that prioritise co-creation over sign-off.
- Language that shifts from "my scope" to "our success."

Equally important is psychological safety, where raising risks is seen as strength, not obstruction. Where ambiguity can be voiced without penalty. Where the team's intelligence is not judged by agreement, but by how it navigates difference.

> *"Real synchronisation means staying in sync,*
> *especially when mistakes happen."*

10.1.4 Why This Section Matters

You can have the best tools, the smartest people, and the most advanced QMS and still fail to deliver if your teams don't move together.

This chapter begins with the most foundational principle: that synchronised minds build synchronised systems. And that collective intelligence is not a nice-to-have. It is the engine of innovation, safety, and speed.

Let us now explore how to embed that synchrony not just into mindsets, but into methods, turning aligned thinking into aligned execution.

10.2 Embedding Synchronisation into Project DNA

It is one thing to aspire to cross-functional collaboration. It is another to embed it so deeply into daily execution that synchronisation becomes second nature.

In high-performing MedTech organisations, alignment isn't something you do at kick-off or fix during fire drills. It is baked into how teams are formed, how decisions are made, and how progress flows every day.

> *"Synchronisation isn't a ceremony. It is a system of working, wired into how problems are framed and how outcomes are owned."*

This section unpacks how synchronisation becomes operational, not just a principle, but a project discipline.

10.2.1 From Handoffs to Integrated Product Teams

Traditional MedTech development mimicked a relay: engineering designs then hands off to QA; QA hands to RA; RA brings in Clinical and Manufacturing. Each function performs, then passes the baton.

But with complex systems, every handoff is a point of friction. Context is lost. Risks emerge too late. And "fixes" are more expensive than alignment ever would have been.

Integrated Product Teams (IPTs) solve this. They are not committees, they are cross-functional engines of execution. IPTs co-own the product from concept to post-market. Not sequentially. Continuously.

A modern IPT includes:

- Engineers and UX designers
- Quality assurance partners
- Regulatory strategists
- Clinical advisors
- Cybersecurity and software leads
- Operations and commercial input

They don't just attend meetings, they shape direction in real time.

> *"Handoffs are what you do when you are not aligned.*
> *Synchronised teams never let go of the baton."*

10.2.2 Aligning Around Shared Outcomes

Synchronisation starts with what teams choose to optimise for. In unsynchronised teams, each function optimises for its own version of success:

- Engineering meets the spec.
- Regulatory hits submission deadlines.
- Quality ensures documentation compliance.

But functional wins don't equal system success.

High-performing teams align around shared outcomes:

- Did the product meet clinical needs?
- Was compliance designed in, or layered on?
- Did we anticipate post-market realities?
- Did the patient experience guide the design?

These are not philosophical questions. They are the basis for how roadmaps are built, trade-offs are made, and success is measured.

"Alignment isn't when everyone agrees.
It is when everyone commits to the same
outcome, even if their inputs differ."

10.2.3 Early Involvement, Continuous Commitment

The most common failure pattern in product development? Late-stage surprises from early-stage exclusions.

Quality wasn't in the room when a critical risk emerged. Regulatory was briefed after prototypes were locked. Clinical surfaced usability issues after final testing.

Synchronised teams invert this. They don't just include functions early, but they keep them engaged throughout. It is not a series of sign-offs. It is a shared cycle of insight, iteration, and ownership.

Continuous cross-functional presence:

- Surfaces risks earlier
- Eliminates costly rework
- Builds empathy between disciplines
- Enables true systems thinking

"You don't build trust through perfect plans. You
build it through shared problem-solving over time."

10.2.4 Why This Matters

Integrated execution is what separates agile-in-name teams from agile-in-practice teams. Synchronisation isn't a philosophy, rather, it is a structure. And when built into the project's DNA, it doesn't slow things down. It speeds the right things up.

In synchronised organisations:

- Delays shrink
- Silos collapse
- Risk becomes a shared radar
- Success becomes shared pride

Let us now turn to how these high-performing teams communicate, coordinate, and stay in sync, not through luck, but through intentional rhythms, platforms, and habits.

10.3 Alignment in Action: Communication and Collaboration That Work

Tools don't synchronise teams. But the right communication culture, supported by purposeful platforms and rituals, absolutely does.

In complex, cross-functional MedTech environments, misalignment rarely happens because people don't care. It happens because they don't see the same version of the problem, or the timing, or the data. Alignment begins not with meetings or messaging tools, but with a shared understanding of what matters, when, and why.

> *"Communication in synchronised teams*
> *isn't noise. It is navigation."*

This section explores how the most agile, aligned organisations turn communication from overhead into advantage.

10.3.1 Cadence Creates Coherence

It is not about how often teams talk. It is about how rhythmically they reconnect.

Synchronised teams establish deliberate communication cadences:

- Daily huddles to surface blockers
- Weekly risk reviews to course-correct
- Sprint retrospectives to reflect and adapt
- Strategic alignment checkpoints across disciplines

These aren't just calendar events. They are alignment rituals: structured moments to recalibrate, reprioritise, and reduce noise.

> *"In complex systems, rhythm drives resilience. Cadence isn't overhead. It is insurance against drift."*

When used well, cadence compresses risk, increases clarity, and helps teams move faster, because they are correcting in real time, not recovering too late.

10.3.2 One Platform, One Truth

Fragmented systems fragment teams. One team uses spreadsheets. Another relies on email threads. Clinical works in SharePoint. Engineering lives in Jira. The result?

- Delayed decisions
- Conflicting versions
- Mistrust in the data

High-performing teams unify on a single source of truth, a digital backbone where everything lives:

- Requirements
- Risk registers
- Test evidence
- Regulatory decisions
- Approval logs

Whether it is Jira, Confluence, Polarion, or a bespoke PLM/QMS system, the value is visibility. When every discipline sees the same data, decision-making becomes faster and better informed.

> *"Clarity is a competitive advantage. One platform, one truth, one team."*

10.3.3 Designing Proximity – Physical or Virtual

Proximity is no longer about office space. It is about access. In hybrid or distributed environments, high-performing teams design for closeness, digitally and emotionally.

- Persistent Slack or Teams channels
- Virtual Kanban boards
- Asynchronous Loom updates
- Real-time collaborative documents

The best teams don't rely on scheduled meetings to stay aligned. They cultivate a flow of micro-conversations that maintain macro-clarity.

> *"Alignment is a posture, not a place. It is how often we connect, not where we sit."*

Intentional digital architecture ensures that geography never becomes a barrier to synchronisation.

10.3.4 Language Fluency Across Disciplines

Many cross-functional breakdowns aren't disagreements, they are simply lost in translation.

- Engineers speak in user stories and sprints.
- Regulatory speaks in clauses and predicates.
- QA speaks in CAPAs and audits.
- Clinical speaks in endpoints and usability studies.

Synchronised teams build cross-functional fluency. They don't expect every team to speak every language, but they create shared dictionaries and mutual understanding.

Tactics that help include:

- Cross-functional teach-ins
- Role-shadowing opportunities
- Shared glossaries in onboarding
- Translators, team leads who bridge disciplines

The result?

- Faster decisions
- Fewer escalations
- Greater respect
- Deeper integration

"Fluency isn't about speaking perfectly.
It is about listening precisely."

10.3.5 Synchronisation as a Communication Culture

Synchronisation isn't just about having better tools. It is about creating an environment where clarity is habitual, inclusion is intentional, and communication is not reactive, but designed.

- Cadence becomes continuity
- Shared systems create shared confidence
- Communication becomes part of quality, not separate from it

Because when teams are connected, not just technologically, but relationally and strategically, complexity doesn't paralyse progress. It fuels it.

Next, we explore the deeper layers of this connection: how culture shapes collaboration, and what it takes to move from surface-level teamwork to a true one-team mentality.

10.4 Breaking Cultural Barriers: Building a One-Team Mentality

You can align timelines, adopt shared tools, even hold cross-functional stand-ups. But if underlying assumptions, stereotypes, and cultural silos remain intact, synchronisation will always fall short.

True cross-functional collaboration isn't a matter of structure. It is a matter of shared identity. High-performing MedTech organisations move beyond coexistence and toward cohesion, creating a culture where collaboration isn't an event, it is an expectation.

They build what might be called a One-Team Mentality, where different disciplines don't just coordinate, they connect.

10.4.1 Dismantling Functional Stereotypes

Every function comes with its own language, priorities, and, over time, its own myths.

Engineers are seen as idealists. Regulatory as blockers. Quality as perfectionists. Clinical as cautious. Marketing as impatient. These caricatures may contain a kernel of truth, but they distort understanding and undermine trust.

Synchronised teams surface and challenge these assumptions, not to shame, but to reframe. When Regulatory pushes back, it is not obstruction, it is protection. When QA asks tough questions, it is not bureaucracy, it is foresight.

The work begins when teams ask: "What assumptions do we make about each other, and what is the story behind them?"

We don't need fewer perspectives. We need fewer assumptions about them.

10.4.2 Building Relationships, Not Just Roles

Collaboration happens between people, not departments.

It accelerates when engineers know the clinical lead by name, when RA is looped into early ideation, and when QA is not just policing quality but co-creating it.

Synchronised teams create deliberate moments for connection:

- Shared design sprints and retrospectives
- Cross-functional onboarding experiences

- Informal peer-learning sessions
- Joint wins celebrated across functions

The result is empathy. And empathy turns compliance into co-ownership and handoffs into handshakes.

In environments where people feel seen, not just for their skillset but for their contribution to the mission, alignment becomes effortless.

10.4.3 Shared Wins, Shared Culture

What organisations choose to celebrate is what they teach teams to repeat.

When wins are claimed in silos, "Engineering delivered" or "RA cleared it," the culture fragments. But when success is narrated as a collective achievement, it reinforces the reality of cross-functional value.

> *"The audit passed clean because QA designed early controls, RA guided documentation, and Engineering integrated risk mitigation from day one."*

This isn't just good storytelling. It is cultural reinforcement. It tells everyone: success here is not about individual excellence, it is about synchronised effort.

What you celebrate, you replicate.

10.4.4 Diversity of Thought as an Asset

Multidisciplinary teams thrive on divergent perspectives. But only when those differences are harnessed, not ignored.

Engineers think in possibilities. Regulatory in boundaries. QA in safeguards. Clinical in user realities. These tensions aren't liabilities, they are design tools.

In disconnected teams, these differences create friction. In synchronised ones, they sharpen insight.

High-functioning teams don't seek uniformity of thought. They design forums for respectful disagreement, scenario mapping, design reviews, and integrated risk sessions where every voice is present and every concern is valued.

Diversity of thinking isn't something to manage. It is something to mine, because when different minds tackle the same mission, innovation becomes richer, faster, and more real.

Synchronisation isn't just about harmonising workflows. It is about elevating relationships. When you build culture intentionally, with trust, curiosity, and mutual respect, collaboration becomes a reflex, not a requirement.

And it is in this cultural soil that the next section takes root: how the best teams use tension, not as a sign of dysfunction, but as a catalyst for clarity and strength.

10.5 Turning Tension into Alignment: Using Conflict Constructively

In MedTech, where lives are on the line and complexity is the norm, conflict isn't a sign of dysfunction, it is a sign that people care. Engineers, regulatory experts, clinicians, and quality leaders bring strong convictions because the stakes demand it. The challenge isn't eliminating tension. It is learning how to use it.

High-performing teams don't avoid disagreement, but they do plan for it. They create environments where friction fuels forward motion and where differences sharpen thinking instead of derailing progress.

> *"When synchronisation is real, conflict*
> *becomes a tool, not a trigger."*

10.5.1 When Conflict Fractures: Silos and Turf Wars

In unsynchronised teams, conflict often feels like a turf war. Engineering defends timelines. Regulatory defends compliance. Quality defends process. Clinical defends patient safety. Each instinct is valid, but when unaligned, they divide rather than strengthen.

The shift happens when teams stop defending positions and start advancing a shared purpose. The goal is not to win an argument, but to serve the patient. Conflict becomes a way to refine, not resist.

10.5.2 Turning Debate into Shared Understanding

High-functioning teams lead with data, not drama. Emotion is acknowledged, but evidence guides decisions. Structured tools – impact analyses, benefit-risk matrices, regulatory mapping – turn subjective debate into objective clarity.

A usability enhancement that raises clinical benefit with minimal added risk can be evaluated transparently, documented effectively, and acted on with confidence. Disagreement, when anchored in data, becomes a pathway to better outcomes.

10.5.3 Preventing Costly Conflicts: Involve Early, Stay Engaged

The costliest conflicts don't come from disagreement, they come from late involvement. When QA flags issues after design is locked, or RA raises concerns after a prototype is built, it is no longer conflict, it is cleanup.

Synchronised teams prevent this by involving every function early and keeping them engaged throughout. Regulatory joins ideation. Clinical shapes features. QA reviews design in real time. The result: fewer surprises and stronger ownership across the board.

10.5.4 The Foundation of Safety: Courageous Clarity

Psychological safety underpins constructive conflict. In strong teams, an engineer can challenge a requirement, RA can flag compliance risks, and Clinical can question design assumptions without fear of blame or dismissal.

Alignment doesn't mean everyone sees the problem the same way. It means everyone commits to solving it together.

10.5.5 From Tension to Energy

The best MedTech teams don't treat tension as a setback. They treat it as energy. And when that energy is channelled with empathy, transparency, and purpose, it becomes momentum, momentum that propels projects forward, not despite complexity, but because of it.

"Conflict, done right, doesn't crack
alignment; it reinforces it."

10.6 The Payoff: Speed, Clarity, and Better Outcomes

When synchronisation moves from concept to practice, the results are immediate and unmistakable. Aligned teams don't just function better, they outperform, outpace, and outlast. Not by pushing harder, but by working smarter, together.

10.6.1 Real Speed: Progress in Parallel

True speed in MedTech doesn't come from cutting corners or skipping steps. It comes from simultaneity, progress happening in parallel, not in sequence.

While R&D prototypes, Regulatory maps approval pathways. While QA builds risk profiles, Clinical integrates user realities. Instead of passing the baton, everyone runs together.

This parallel progress reduces handoffs, shortens delays, and prevents problems before they grow. What once took months of back-and-forth becomes a fluid, shared journey. Synchronised teams make momentum look effortless, but it is rigorously designed and deeply earned.

10.6.2 Quality as a Shared Discipline

In synchronised teams, quality stops being a department and becomes a discipline shared across functions.

When RA, QA, Clinical, and Engineering engage from the outset:

- Design flaws are caught early.
- Usability is validated meaningfully.
- Compliance is built into every feature.

The payoff: fewer late-stage surprises, fewer regulatory escalations, and fewer post-launch regrets.

It is not about adding more layers of review, it is about building more layers of awareness: early, ongoing, and collective.

10.6.3 Culture Shift: From Compliance to Commitment

Beyond processes and metrics, synchronisation transforms how teams feel. Ownership deepens. Silos dissolve.

- QA is no longer a gatekeeper, but a guardian.
- Regulatory is not an afterthought, but an enabler.
- Clinical is not a checkpoint, but the voice of the user.

When every team member sees their work reflected in the success of the whole, morale shifts from compliance to commitment. Energy is no longer drained by firefighting, it is focused on forward-building.

Retention improves. Creativity sharpens. Teams don't burn out, they buy in.

Synchronised teams move not just with precision, but with pride. They deliver not only faster products, but safer innovations. And they do it with a sense of cohesion that patients, regulators, and partners can feel.

10.7 The Human Multiplier: Why People Are the Ultimate Key

Tools can align processes. Frameworks can structure workflows. But only people can synchronise a system. The real engine behind high-performing, cross-functional teams isn't technology, it is trust, shared purpose, and relational intelligence.

10.7.1 Culture as the Conductor

Culture isn't the backdrop, it is the conductor. It decides whether synchronisation is aspirational or fully alive.

In synchronised cultures, collaboration isn't exceptional, it is expected. Raising a concern is seen as an act of leadership, not disruption. Success isn't hoarded within silos but celebrated across functions. And when pressure mounts, it is culture, not process, that holds the team together.

10.7.2 Leadership as the Catalyst

Leaders model the behaviours that matter: transparency, bridge-building, curiosity across disciplines. They make psychological safety a strategic asset. In such environments, quality is co-created, not imposed. Innovation flows from shared understanding, not endless alignment meetings.

10.7.3 Relationships Over Roles

Synchronisation isn't built by job titles or org charts, rather it is built through human connection.

- Engineers and clinicians who trust each other resolve ambiguity faster.
- QA and RA professionals who understand intent spot and solve issues earlier.

When respect is the foundation, synchrony becomes second nature.

10.7.4 People as the Multiplier

The future isn't about rigid control, it is about intentional collaboration. As MedTech grows more complex, people become the multiplier, the force that transforms technical teams into innovation ecosystems.

The greatest accelerant to your pipeline isn't a tool or template. It is aligned people moving with clarity and conviction. Competitive advantage isn't only how fast you deliver, but how well your people deliver together.

Synchronisation doesn't just produce better outcomes. It builds better organisations. And in a field where lives are at stake, that alignment is not optional, it is a leadership imperative.

10.7.5 A Call to Align

Systems don't synchronise themselves, people do. Use these questions to spark reflection and alignment in your teams:

- Are your cross-functional teams truly collaborating, or just coordinating in silos?
- What tools or practices could build greater transparency, rhythm, and shared understanding?
- Where cultural or language barriers are silently slowing progress and are they ever addressed directly?
- How do your teams handle conflict? Does it fuel better outcomes or avoidable delays?
- Are your people empowered to own alignment, or waiting for someone else to create it?

As we move into Chapter 11, we shift from the human rhythm of collaboration to the intelligence behind performance. Synchronised teams need more than alignment, they need visibility. Up next: how to harness predictive metrics, simplify measurement, and turn data into decisive, intelligent action.

METRICS THAT MATTER – PREDICTIVE QUALITY & PERFORMANCE INTELLIGENCE

W hen a pilot navigates through cloud cover, they don't rely on instinct, they rely on instruments. Real-time metrics like altitude, heading, and velocity translate uncertainty into confidence. It is not any single gauge, but how those signals work together and are interpreted in context that ensures a safe landing.

The same is true in MedTech. Metrics aren't just dashboards or compliance checks, rather they are decision enablers. They transform blind spots into clarity and drive action with speed and confidence. As products become smarter and expectations more real-time, the challenge isn't measuring more, it is measuring what matters.

In this data-saturated environment, insight has become the rarest commodity. Many organisations are overwhelmed by charts and KPIs that inform meetings but not decisions. The opportunity is to shift from volume to value, from passive reporting to performance intelligence. From lagging hindsight to leading foresight.

This chapter explores that shift. How high-performing teams design metrics that shape behaviour, surface early signals, and enable action. We will look at choosing fewer, smarter indicators, harnessing predictive analytics, and transforming dashboards into living systems of improvement.

In synchronised organisations, metrics aren't someone else's job. They are a shared language, a way to align, decide, and build trust. Because in regulated innovation, readiness often depends on how well we listen to what the data is already telling us.

This isn't about tracking everything. It is about knowing what to track, why it matters, and how to turn numbers into narrative and narrative into momentum.

11.1 From Data to Insight

In fast-moving MedTech, data isn't a by-product, it is a core input. It guides decisions, aligns functions, and enables early intervention. But the shift isn't just in tools, it is in mindset: from seeing data as a report card to using it as a compass.

11.1.1 The New Role of Data

Traditionally, teams acted first and analysed later. But now, analytics and execution must move in lockstep. Whether it is a yield tracking on the line or test coverage in software, data is becoming a real-time signal system, shaping work as it happens.

This is crucial in an era of rising complexity and shrinking timelines. Leading organisations treat data as operational fuel, empowering faster decisions, earlier adjustments, and continuous alignment.

11.1.2 From Lagging to Leading

Legacy systems rely on lagging indicators: complaints, deviations, audit findings. They are necessary, but insufficient.

Today's leaders balance them with leading indicators, early signals that risk may be forming. Think of:

- Timeliness of design reviews
- Early yield trends
- Training effectiveness
- CAPA closure velocity
- Field signal anomalies

Leading metrics allow action before harm occurs. They don't replace lagging data, they complete the picture. Acting early demands a cultural shift: learning to trust early signals and moving without waiting for crisis.

11.1.3 Predictive Quality

Organisations are drowning in dashboards but starving for insight. The issue isn't data, it is clarity.

Predictive quality applies statistics and machine learning to detect subtle patterns: process drift, weak links, or systemic noise. For example, small deviations flagged a design-for-manufacturing issue that, once corrected, prevented a six-month delay.

It is not about complexity but about foresight. Even basic tools like trend lines can reveal hidden risks if framed right. The goal: data that triggers timely action by the right people.

11.1.4 Leadership and Culture

Metrics only matter if they move people, and that starts with leadership.

Leaders must create a culture where data is reviewed routinely and acted upon confidently. This includes:

- Integrating metrics into daily and weekly cadences
- Holding teams accountable for prevention, not just correction
- Promoting transparency, even when data is uncomfortable

Importantly, data must not be weaponised. It should function as a mirror, reflecting, not blaming. In psychologically safe environments, teams engage with curiosity, not defensiveness. They explore the "why" behind the trend and act on it.

Predictive quality isn't a tool. It is a mindset. And it becomes real when data flows across silos, informs timely decisions, and supports the one goal that matters most: safe, effective outcomes for patients, not eventually, but consistently.

11.2 Choosing the Right Metrics: Less Is More

In a data-saturated environment, the temptation to measure everything is high. But high-performing MedTech organisations don't aim for quantity, they aim for clarity. Metrics should not impress, they should inform. The goal is not to fill dashboards but to guide action.

Metrics become powerful when they shape decisions, provoke action, and reflect what truly matters. If they don't, they are noise, and noise confuses more than it clarifies.

11.2.1 Focus on What Matters

A flood of metrics often leads to diluted focus and delayed decisions. Effective teams align around a small set of high-value indicators, ones with a direct line to outcomes, not just outputs.

The guiding question isn't "What can we measure?" It is "What decision does this enable?"

Smart metrics connect strategy to execution. They sharpen conversations and align teams toward impact, not activity.

As Crosby taught us:

> *"The highest quality is not measured by how well we fix defects, but by how well we prevent them from occurring in the first place."*

11.2.2 Balancing Lagging and Leading Indicators

Lagging indicators, like audit findings or defect rates, tell us what went wrong. They are essential but retrospective. Leading indicators, by contrast, offer foresight as risk signals before harm materialises.

Examples of leading indicators:

- Delayed CAPA initiation
- Dips in training completion
- Variability in early-stage yields
- Slips in internal audit schedules

A balanced system combines both: hindsight for learning and foresight for prevention.

11.2.3 Cutting Through the Noise

Vanity metrics, like document counts or meeting volumes, may look productive but rarely guide improvement. Effective metrics are:

- Relevant: Linked to strategy and risk
- Actionable: Trigger decisions or interventions
- Timely: Available when action is possible
- Contextual: Interpreted alongside related signals

If a metric doesn't answer, "What should we do differently?" it is decoration, not insight.

As the timeless wisdom reminds us:

> *"Not everything that can be counted counts, and not everything that counts can be counted."*

A synchronised, intelligent system must apply one golden rule:

> *"Measure what makes you better, not what makes you look better."*

11.2.4 Less Is More: The 10/3/1 Rule

A practical model used by many MedTech leaders:

- 10 key metrics across the value chain
- 3 core metrics per function
- 1 defining metric per strategic priority

This approach creates shared focus. It embeds clarity into daily operations and ensures that metrics are reviewed and acted upon, not just displayed.

11.2.5 Early Signals: A Case in Action

A start-up launched a Bluetooth-enabled auto-injector. While compliance metrics looked fine, customer support noticed an uptick in connectivity questions. A deeper look revealed a firmware glitch on specific mobile OS versions. A quick update prevented a potential recall.

The insight: small anomalies often precede large issues. Great teams listen to whispers before they become alarms.

11.2.6 Designing Metrics with Purpose

Each metric should exist for a reason. Ask:

- What behaviour will this drive?
- What decision will it support?
- What risk does it help us mitigate?

Metrics are not external scorecards. They are internal compasses, steering action, not just tracking it.

11.2.7 Shared Ownership, Sharper Insight

Metrics designed in isolation often lack trust. But when teams co-define what is measured and why, they take ownership and accountability follows.

Cross-functional metrics, owned by both QA and R&D, or Clinical and RA, build alignment and reduce blame. Dashboards become more than performance monitors, they become conversation starters and improvement accelerators.

11.2.8 Final Thought: Precision Over Proliferation

You don't need more metrics. You need better ones. The right set reveals what matters, drives the right behaviours, and simplifies decisions under complexity.

In a high-stakes, high-velocity industry like MedTech, clarity is not a luxury, it is a strategic asset.

11.3 Predictive Analytics: Anticipating Issues

Traditional MedTech quality systems have been largely reactive, responding to deviations, complaints, and CAPAs after the fact. But in today's faster, more complex environment, lag is risk. The future belongs to those who can anticipate, not just react.

Predictive analytics enables this shift by spotting trends, surfacing risks, and prompting action early. It is not just a data capability, it is a cultural evolution from hindsight to foresight.

11.3.1 Trend Awareness – Seeing the Early Drift

Prediction begins with pattern recognition, not AI. Subtle changes in recurring metrics often precede major issues: slight dips in yield, repeated near misses, increased rework.

One company spotted a rise in micro-leaks, each individually compliant, but collectively signalling a materials issue. Addressed early, this avoided downstream rework and potential complaints.

Trend analysis doesn't require sophisticated tools, just attentiveness. Direction often matters more than the number itself.

*"If you can spot the clouds forming, you
don't have to get caught in the storm."*

11.3.2 Smarter Devices, Smarter Insights

Connected MedTech devices generate continuous performance data, opening new possibilities for field-based prediction.

Imagine a smart pump detecting a voltage drift. Alone, it is noise. But across thousands of units, it signals emerging failure. Pre-emptive firmware updates replace recall risk with routine maintenance.

Another firm used telemetry to correlate Bluetooth module issues with app disconnections, solving it with a targeted software patch before formal complaints began.

Predictive insight transforms field data into frontline prevention.

11.3.3 AI in Quality – Potential with Guardrails

AI can amplify prediction, detecting patterns too complex for humans to spot. But in regulated settings, AI must be explainable, validated, and governed. The promise is powerful – faster signal detection, smarter triage, and better prioritisation – but only if risk is managed.

Best practice includes:

- Using explainable AI over black-box models
- Validating outputs with real-world data
- Keeping humans in the loop for final decisions
- Ensuring traceability for regulatory scrutiny

AI should enhance judgment, not replace it.

11.3.4 Building Early Warning Systems

Leading organisations embed predictive triggers into operational workflows. Examples include:

- Escalation when yields drop >10%
- Complaint clustering triggers cross-functional triage
- Repeated audit findings initiate targeted retraining

These triggers don't add bureaucracy, they build agility. By detecting weak signals early, teams prevent stronger consequences later.

11.3.5 Earning Trust Through Insight

Predictive analytics earns trust when it leads to timely, transparent action. The result:

- Fewer surprise CAPAs
- Faster time-to-resolution
- More confident regulatory interactions
- Stronger cross-functional alignment

But trust must be earned. This means explaining the logic behind predictions, validating models regularly, and acting responsibly.

In MedTech, where safety, compliance, and innovation intersect, prediction is no longer a luxury. It is a leadership imperative.

11.4 Closing the Loop: Using Metrics for Decisions and Improvement

Metrics aren't valuable until they move from dashboards to decisions. In leading MedTech organisations, data isn't passively reviewed, it is actively used to shape choices, guide priorities, and fuel continuous improvement. Closing the loop means turning observation into action.

11.4.1 From Measurement to Movement

Metrics must drive momentum. If they merely describe performance, they are decorative. When embedded into daily rhythms and tied to real decisions, they shift from reports to levers, helping teams adjust course in real time.

The goal is not just to track what is happening but also to influence what happens next.

11.4.2 Embedding Metrics into Operating Rhythms

Metrics shouldn't live in annual reviews, rather they must live in daily work:

- Stand-ups spotlight yield or defect trends, with action logged immediately
- Weekly reviews surface CAPA cycle delays or rising complaint rates
- Monthly syncs align QA, RA, Ops, and Clinical on systemic risks
- Quarterly leadership reviews use trends to inform strategic pivots

When metrics are consistently reviewed and acted on, they become part of the team's muscle memory, not just management's view.

"Metrics are not the report card after the race;
they are the compass during the journey."

11.4.3 Clear Ownership, Clear Action

Every metric needs a name, not a label, but an accountable owner. Ownership turns abstract numbers into concrete priorities. Without it, even the most insightful metric fades into the background.

Whether it is complaint response times, software build stability, or CAPA closure rates, someone must own the improvement story and be empowered to lead it.

11.4.4 Designing Metrics That Drive the Right Behaviour

Poorly designed metrics distort. When a number becomes the target, it can drive gaming or shortcutting, what Goodhart's Law warns against.

A team might cut test time to meet velocity targets, only to miss critical defects. Strong metrics balance speed with depth, clarity with context. They are reviewed regularly, challenged openly, and evolve with purpose.

The question is not "Are we green?" but "Are we improving in the right direction?"

11.4.5 Creating a Learning Culture, Not a Blaming One

Metrics should never be weapons. If red triggers blame, teams hide problems. If it triggers curiosity, teams surface problems early.

One company introduced "metric retrospectives," judgment-free reviews focused on trends and learning. The shift was cultural: data became a shared story, not a scorecard.

In mature teams, metrics are mirrors, tools for self-awareness and growth.

> *"When a measure becomes a target, it*
> *ceases to be a good measure."*

11.4.6 From Reporting to Real-Time Leadership

In the most agile organisations, metrics don't trickle up. They pulse across. Teams see, decide, and act in shorter cycles, because leadership trusts them with the data, and they trust themselves with the responsibility.

That is what performance intelligence truly means: real-time awareness, shared ownership, and the confidence to improve without being told.

As one MedTech leader put it: "We stopped managing by review. We started leading by readiness."

> *"Metrics must be designed for truth, not just targets."*

11.5 Tools for Performance Intelligence: Beyond Excel

For years, spreadsheets were the backbone of MedTech reporting. But static tools can't keep pace with today's complexity. Modern performance intelligence requires more than tracking, it requires tools that connect data, context, and decisions in real time.

The goal is no longer just to collect data. It is to translate it into visibility, enablement, and action.

11.5.1 From Data Storage to Decision Support

Legacy tools gather data. Intelligent tools drive decisions. Modern platforms unify data streams from QMS, MES, CRM, and ERP, converting raw numbers into shared situational awareness.

This shift turns data from passive archives into active engines of alignment, helping cross-functional teams respond faster and more effectively.

11.5.2 Real-Time Dashboards as Operational Catalysts

Dashboards today are more than visual summaries, rather they are operational control panels. When quality, regulatory, and operations teams access shared, real-time dashboards, issue detection becomes collaborative and response becomes immediate.

In one company, a spike in complaint signals across global sites triggered same-day investigation, before it reached customers or regulators. Action wasn't delayed by reporting. It was led by visibility.

"Dashboards aren't for watching the game from the stands – they are for playing it smarter on the field."

11.5.3 Empowering Frontline Teams with Embedded Analytics

Performance tools should empower the people closest to the work. Embedded analytics let operators, engineers, and support teams monitor their own metrics and act early.

A software team tracking test coverage mid-sprint or a manufacturing cell adjusting process parameters after seeing a yield drop – this is where metrics gain meaning. Not in boardrooms, but on the floor.

11.5.4 Benchmarking That Adds Context, Not Pressure

Comparing metrics across sites or with external peers can spark improvement, if done wisely. Benchmarking must be contextual, not competitive.

A 60-day CAPA closure time may look strong until peers average 30, but what is the risk profile, market scope, or maturity stage? Benchmarking should guide aspiration, not enforce imitation.

Done right, it is less about comparison and more about calibration: what is possible and what is worth aiming for.

11.5.5 Connecting Systems to Eliminate Blind Spots

Disconnected tools create data silos and decision delays. Leading organisations invest in integration, linking design controls, quality events, and post-market data for seamless insight.

When complaint trends connect directly to design specs or training records to audit readiness, decisions become faster and more confident.

Integration is not a tech project, it is a strategy to align data flow with decision flow.

11.5.6 Real-World Impact of Intelligent Tools

One global firm used integrated dashboards to flag rising scrap and CAPA trends in a single plant. A supplier issue was identified, resolved, and contained before it affected customers. The technology didn't solve the problem, but it made the problem visible, fast.

The lesson: tools don't improve performance. People do, when they are equipped with clarity.

11.5.7 Final Thought: Visibility Fuels Velocity

Excel isn't the enemy, inertia is. To lead in modern MedTech, organisations must move beyond isolated spreadsheets to systems that illuminate patterns, empower people, and support smarter, faster decisions.

Performance intelligence is not about digital transformation alone. It is about creating clarity, building trust, and turning insight into impact.

11.6 Empowering Teams: Frontline Ownership of Metrics

In many MedTech organisations, metrics live in dashboards for leadership. But the real power of performance intelligence is unleashed when metrics reach the people closest to the work, those on the frontline. Ownership turns numbers into action and data into daily decisions.

11.6.1 From Measurement to Meaningful Ownership

Metrics should not be something done to teams, but something done with them. When engineers, operators, and clinical teams understand, trust, and influence the metrics they track, they shift from passive observers to active stewards of performance.

Ownership fosters accountability, motivation, and faster problem-solving, not because teams are told what matters, but because they believe it does.

11.6.2 Visibility Where It Counts

Frontline teams can't respond to what they can't see. Data must be accessible, relevant, and reviewed regularly, in the right place, at the right time.

This might mean:

- Line operators tracking defect trends and adjusting parameters
- QA leads reviewing open CAPAs during team huddles
- Software teams monitoring real-time rework rates during sprints

These aren't extra tasks, they are embedded habits that keep improvement constant.

11.6.3 Decision-Making at the Point of Impact

When teams see live data, they can act immediately. They don't need escalation pathways, but they do need clarity and empowerment.

For example:

- A spike in unresolved service tickets prompts real-time resourcing adjustments
- A dip in yield triggers on-the-spot calibration
- An uptick in device disconnects sparks collaborative triage before it becomes a complaint

To support this, organisations define clear rules of engagement: who owns the metric, what the thresholds are, and how escalation works.

"In great organisations, data does not
silence people - it sharpens them."

11.6.4 Red as a Signal, Not a Failure

In high-performing cultures, metrics are feedback, not judgment. A "red" KPI doesn't trigger blame, it triggers curiosity. It starts a conversation: What is the root cause? What can we learn?

When leaders respond to issues with support instead of stress, transparency grows. Teams speak up earlier, and improvement accelerates.

Metrics in a high-trust culture shift from oversight to insight and from pressure to possibility.

11.6.5 From Reporting to Engagement: A Real-World Story

A MedTech company's service team faced persistent backlogs. Management dashboards had little effect, until the teams themselves took control.

Live dashboards were introduced at the team level. Daily huddles became data-driven. Frontline agents identified friction points, restructured workflows, and proposed tech fixes. Within weeks:

- Resolution time dropped 30%
- First-contact resolution improved
- Team morale spiked

One team member said it best:

"We stopped being measured and started measuring, and that changed everything."

11.6.6 Institutionalising Ownership

True ownership becomes cultural when it is built into the fabric of how teams work:

- Teams co-define the metrics they use
- Metric reviews are part of team rituals, not just management reviews
- Recognition is given not just for results, but for insight and initiative

Ownership is about agency. When teams know the numbers, trust the data, and feel empowered to act, performance becomes a shared pursuit, not a top-down mandate.

11.6.7 Final Thought: The Frontline as a Force Multiplier

When metrics are owned at the frontline, change happens where it matters most, where the work is done, and where patients are ultimately impacted. Empowered teams don't wait to be told what is broken. They detect, decide, and deliver. Every day.

Ownership is the heartbeat of a high-velocity, high-trust organisation. And in MedTech, it is not just a culture shift, it is a strategic advantage.

11.7 Outcomes: Proactive and Intelligent Organisations

When metrics are embedded deeply in operations, they do more than inform, they transform. High-performing MedTech organisations use data not just to monitor what happened, but to shape what is next. The result is a culture that is proactive, responsive, and continuously improving.

11.7.1 Metrics as Operating Muscle, Not Monthly Ritual

In intelligent organisations, metrics become second nature. They are part of how teams work, not an add-on. Engineers review design metrics before planning iterations. QA uses real-time yield data to adjust processes. Clinical teams reference usability trends to fine-tune requirements.

This fluency builds discipline. The habit of using data daily creates resilience and precision, not just during audits, but across the lifecycle.

11.7.2 From Firefighting to Forward-Thinking

With strong metrics in place, problems are identified early, before they cascade into nonconformances, CAPAs, or complaints. Teams shift from reacting to resolving.

Instead of managing late-stage issues, they adjust inputs upstream. Instead of scrambling during audits, they already know where the gaps are and have addressed them. In these environments, firefighting is the exception, not the norm.

11.7.3 Aligned Decisions and Faster Action

Data reduces debate. When teams rely on shared metrics, decisions become faster and more objective. Trade-offs are made with clarity, not opinion. Teams stop asking "who is right?" and start asking "what does the data show?"

This fosters trust, eliminates ambiguity, and streamlines collaboration across functions, a critical advantage in complex, regulated environments.

11.7.4 Understanding Variation to Guide Smarter Responses

A mature metrics culture doesn't just chase changes in numbers, it also interprets them wisely. Teams learn to differentiate between common cause variation (normal system noise) and special cause variation (true signal).

This insight prevents overreaction and underreaction alike. It also leads to smarter, more surgical interventions: saving time, effort, and resources.

11.7.5 Metrics That Drive Confidence

Well-used metrics don't just improve performance, they also increase confidence across the organisation:

- Frontline teams feel ownership, not oversight
- Quality leaders operate from insight, not instinct
- Executives lead with transparency, not pressure

This shows up in measurable ways:

- Fewer late-stage surprises and quality escapes
- Reduced cost of poor quality
- Better audit readiness
- Stronger market responsiveness

As one senior leader put it:

> *"We stopped tracking activity. We started measuring progress, and that changed how we lead."*

11.8 Intelligent Growth

The future of MedTech doesn't belong to the fastest movers, it belongs to the most intelligent ones. Growth rooted in visibility, foresight, and alignment isn't just sustainable, it is transformative. And at the heart of this transformation is not more metrics, but smarter ones.

> *"In the hands of the wise, metrics are not numbers to fear - they are stories waiting to be understood."*

11.8.1 Metrics as a Common Language of Progress

Metrics aren't just tools for tracking performance, they are how teams align, how strategy is translated into action, and how progress is communicated without noise. When designed intentionally and reviewed regularly, metrics unify diverse teams around a shared understanding of what matters and why.

Used well, metrics help teams ask better questions, make faster decisions, and stay focused on patient-centred outcomes.

11.8.2 Focus Is a Leadership Advantage

In an age where dashboards overflow and alerts compete for attention, strategic focus is no longer a luxury, it is a leadership requirement. High-performing organisations resist the urge to track everything. They filter constantly: Is this metric still relevant? Is it driving behaviour or just creating noise?

The most effective metrics are continuously refined to reflect shifts in goals, risks, and capabilities. Focus isn't about less metrics, rather it is about sharper ones.

11.8.3 Insight as the New Speed

True speed in MedTech comes not from rushing but from reducing rework, eliminating blind spots, and acting on insight early. It is the engineering team that spots a drop in first-pass yield and adjusts before defects multiply. The clinical team that links declining app engagement to usability friction and fixes it before complaints escalate.

Insight, not activity, becomes the real driver of speed and responsiveness.

11.8.4 Metrics in Action Across the Value Chain

When metrics are fully embedded, quality becomes proactive. Regulatory timelines tighten because document readiness is predictable. Post-market surveillance is paired with real-time field data. Design teams use usability feedback to iterate before verification, not after. Manufacturing sites detect yield shifts and respond autonomously.

In these systems, teams aren't preparing for audits, they are ready by default. The organisation moves as one: informed, aligned, and agile.

11.8.5 From Measurement to Maturity

When metrics are misunderstood, they create friction. When built with care and used with courage, they accelerate maturity. They enable teams to lead with facts, to course-correct with confidence, and to improve continuously, not just because they are told to, but because they choose to.

Smart metrics don't just track what matters. They reinforce what matters. And when paired with trust, clarity, and a learning culture, they become the engine of intelligent growth.

This is not just a call to reflect. It is a call to measure what matters. Metrics should guide, not distract. Ask yourself and your teams bold, practical questions:

- Are the metrics you track driving meaningful action, or simply reporting what is easy to measure?
- How often do your teams use data proactively to predict and prevent issues, rather than react to them?
- Is your performance intelligence system designed for speed, clarity, and learning, or buried in manual, disconnected tools?
- Are frontline teams empowered to understand, trust, and act on the data, or is it owned by a select few?

As we close this chapter on performance intelligence and the power of meaningful metrics, we now turn our focus to Chapter 12: Beyond Compliance – Turning Audits, V&V, and Docs into Value Drivers.

The next chapter challenges a long-held assumption, that audits, validations, and documentation are merely obligations to survive. Instead, we will explore how leading MedTech teams are reimagining these activities as opportunities to accelerate learning, strengthen design, and build trust. Because when compliance is approached as a source of value, not just a box to check, quality becomes not a cost, but a catalyst.

CHAPTER 12

BEYOND COMPLIANCE – TURNING AUDITS, V&V, AND DOCS INTO VALUE DRIVERS

For many organisations, the words "audit," "verification," "validation," and "documentation" still trigger a familiar reflex: anxiety, checklist fatigue, and the rush to comply. These activities are often treated as burdens, overhead to manage or survive on the way to approval. But what if they weren't hurdles to clear, but catalysts for better performance, faster innovation, and smarter decisions?

This chapter challenges the traditional view. It reframes compliance activities not as the cost of doing business, but as untapped assets, lenses through which organisations can learn, improve, and scale. Done well, audits become strategic intelligence. V&V becomes a driver of innovation. Documentation becomes institutional memory, driving clarity and accelerating decisions.

The shift is not in what we do, but how we see it. Excellence is not found by doing the minimum required, it is unlocked by treating every regulatory requirement as an invitation to be better.

"When you see compliance as a floor, you survive.
When you see it as a launchpad, you soar."

This is not about working harder. It is about seeing smarter. In the following pages, we explore how forward-thinking teams are turning compliance into competitive advantage and how your organisation can do the same.

12.1 Redefining the Role of Audits

12.1.1 From Policing to Performance Insight

Audits have long been viewed as events to survive, regulatory spot-checks that spark reactive behaviour, rushed documentation, and temporary fixes. The result is stress without sustainability. But this fear-based approach erodes trust and wastes time.

Leading organisations now treat audits as opportunities: structured moments to test system health, reveal blind spots, and improve proactively. Audits become less about passing and more about learning. Less about inspection, more about insight.

12.1.2 The Cost of a Defensive Mindset

When audits are treated as interruptions, teams focus on appearances rather than accuracy. Findings become fire drills. CAPAs become box-ticking exercises. Repeat issues re-emerge because root causes go unaddressed.

This approach consumes valuable time and lowers transparency. It creates a culture where teams aim to "get through" the audit, rather than improve through it.

12.1.3 Audits as System Health Checks

Progressive organisations see audits as mirrors, reflecting how well the quality system functions in practice not just on paper. They ask:

- Are we delivering quality consistently?
- Do controls work under real conditions?
- Is our culture aligned with compliance, or merely compliant?

Internal audits become rehearsal for external scrutiny, structured opportunities to test resilience, not just readiness.

> *"An audit isn't a search for flaws – it is*
> *a search for untapped potential."*

12.1.4 Embedding Audit Value Across the Business

Audits become more valuable when they move beyond the Quality silo. Leading companies:

- Co-plan audits with business units.
- Focus scope on risk, change, and recent trends.
- Involve cross-functional leaders in execution and resolution

This turns audits into business assets, helping inform strategy, not just satisfy regulators. It also shifts perception: from "Quality auditing Quality" to "the business auditing itself."

12.1.5 Case Example: From Audit Burden to Business Insight

A diagnostics firm once viewed audits as painful formalities. CAPAs took 90+ days to close. Repeat issues were common.

After reframing audits as improvement sprints, they tracked systemic insights, closure velocity, and recurrence rates. Cross-functional leaders joined root cause reviews. Within a year:

- CAPA closure time fell by 40%
- Repeat findings dropped by 60%
- Audit sentiment shifted from burden to benefit

As one leader put it:

> *"It stopped being about compliance and*
> *started being about confidence."*

12.1.6 From Findings to Follow-Through

Audit effectiveness isn't about detection, it is about resolution. Best-in-class teams:

- Assign clear ownership and due dates.
- Track CAPA effectiveness, not just closure.
- Review long-term impact and adoption rates.

Some organisations revisit audit findings six months later to assess whether the solution stuck. That is not just closure, that is learning.

12.1.7 Audits as Culture Signals

How audits are conducted speaks volumes. Top-down, fear-driven audits lead to defensive behaviours. Constructive, inclusive audits foster transparency and trust.

Leaders set the tone. When executives show up to closing meetings, ask questions, and support action, audits become a tool for transformation, not tension.

> *"External audits are not just checkpoints. They are mirrors — revealing what we have missed and reflecting how far we have come."*

12.1.8 Reframing for the Future

As MedTech becomes more digital, distributed, and AI-enabled, audits must evolve too. Smart organisations are already asking:

- Are we audit-ready for AI-driven submissions?
- Do our audits reflect the complexity of global supply chains?
- Are insights from audits feeding into risk management and design strategy?

The most synchronised organisations no longer ask, "Did we pass?" Instead, they ask, "What did we learn, and how will it make us better?"

Audits, done well, are not interruptions. They are instruments of evolution, tools to test, adapt, and grow. They don't slow you down. They sharpen your edge.

12.2 Audits as Strategic Intelligence: More Than Pass/Fail

12.2.1 From Scorekeeping to Sensemaking

In traditional quality systems, audits are treated like pass/fail exams. Did we comply, or didn't we? This binary view overlooks the deeper value of audits as tools for sensemaking. Leading MedTech organisations are redefining audits as engines of intelligence, mechanisms to uncover weak signals, understand system behaviours, and improve enterprise learning.

The new question isn't "Did we pass?" It is "What did the audit teach us?"

In this paradigm, audits are no longer compliance checks conducted in isolation. They become embedded in how an organisation learns: locally, cross-functionally, and system-wide.

12.2.2 Audit Maturity – A Four-Level Lens

Audit maturity evolves through distinct levels, each unlocking greater insight and organisational value:

Maturity Level	Audit Focus	Primary Outcome
Level 1	Compliance checklist	Pass/fail result
Level 2	Risk-aligned planning	Process conformance
Level 3	Systems thinking	Systemic opportunity identification
Level 4	Integrated foresight	Enterprise learning and improvement

At Level 4, audits are no longer Quality-owned routines. They are enterprise-wide drivers, integrated into risk management, design reviews, and strategic decisions.

12.2.3 Prioritising What to Audit – Intelligence Over Routine

Rather than rotate audits uniformly, leading teams apply intelligent prioritisation, asking:

- Where does patient risk converge with process complexity?
- Which functions are experiencing high change or turnover?
- What signals (complaints, deviations, trending CAPAs) suggest weak points?
- Where are ownership boundaries unclear or contested?

One global MedTech company used a heatmap combining product complaints, design changes, and CAPA activity to reprioritise audits. This surfaced a weak spot in design verification, leading to a redesign that prevented a costly recall, an insight no checklist would have flagged.

12.2.4 From Auditor to Investigator

Auditors in advanced systems move beyond compliance verification. They act as investigators, seeking to understand why issues arise and what they reveal.

Effective auditors ask:

- What patterns emerge across recent NCs?
- What system gaps, not just procedural missteps, are driving risk?
- Are behaviours, training, or handoffs contributing to these trends?

They combine observations with broader context – employee feedback, training metrics, and tool usage – to paint a fuller picture. The goal is not to count findings but to connect them.

One firm began clustering audit data across global sites. A repeated failure in process change communication emerged, invisible in individual reports but obvious when viewed collectively. That insight drove a systemic change that improved product readiness across four regions.

12.2.5 Real-World Example – Product Intelligence from an Audit

During a routine audit of a post-market surveillance process, a mid-sized implantable device company noticed something unexpected. Complaint reporting was compliant, but field service technicians were not logging subtle product anomalies, not due to neglect, but because they didn't recognise them as reportable.

This wasn't a nonconformance, it was a signal. The organisation redesigned intake scripts, retrained technicians, and improved early detection of low-frequency events. Months later, this prevented a potential Class II recall. That is not just passing an audit, that is transforming one.

12.2.6 Creating a Networked Audit Brain

In high-performing companies, audit insights are shared, analysed, and repurposed. Audits are not stand-alone events, they are part of a quality intelligence system. Key enablers include:

- Digital audit platforms with real-time dashboards and risk-based scheduling
- Cross-site calibration for rating consistency
- Audit knowledge bases that evolve with every cycle

Findings also feed into other systems:

- Recurrent NCs → Updated training
- Design transfer gaps → Revised V&V plans
- Supplier issues → Escalated Supplier Relationship Management oversight
- Weak root cause data → Refreshed CAPA methodology

Audits aren't endpoints, they are inputs to continuous improvement, traceable and measurable.

12.2.7 Bringing Audits to the Strategy Table

To unlock full value, audit outputs must reach decision-makers. Best-in-class companies integrate key insights into:

- Quality Councils: for systemic issue alignment
- Design Reviews: for linking findings to product maturity
- Executive Dashboards: for demonstrating foresight, not just compliance

Leadership doesn't ask, "How many findings did we have?" They ask, "What do the findings reveal about our ability to scale, improve, and innovate?"

12.2.8 Final Thought – The Audit as a Force Multiplier

Audits no longer need to be feared, tolerated, or hidden. When done right, they become powerful tools of collective intelligence. They help organisations detect weak signals, connect dots, and build foresight.

In the synchronised MedTech enterprise, the audit is not a scorecard. It is a lens, a way of seeing the organisation more clearly and evolving more deliberately.

The smartest companies don't just pass audits. They mine them for wisdom. They use them to ask better questions, make faster decisions, and build more resilient systems, long before regulators arrive.

12.3 V&V Reimagined: From Regulatory Hurdle to Strategic Catalyst

12.3.1 From Checklist to Confidence

In many MedTech organisations, Verification and Validation (V&V) are still treated as end-of-line hurdles, checkboxes to cross for submission, rather than strategic tools for insight and innovation. This mindset limits their potential.

But in high-performing organisations, V&V is no longer about satisfying regulators, it is about building confidence. Confidence that the product works, under real conditions, for the intended users. Confidence that safety, usability, and performance are not assumptions, but demonstrated truths.

> *"V&V is not about passing a test, it is about proving what matters, before patients ever find out we didn't."*

12.3.2 Embed Early, Design Better

Traditional V&V begins too late, after key decisions are made, when changes are costly. This often leads to rework, bottlenecks, and avoidable risk exposure.

Modern MedTech flips the model. V&V starts early, shaping decisions, not just testing them. Validation engineers sit at the design table. Requirements are written to be testable. Risks are mapped before architecture is finalised. Prototypes and simulations are used to challenge assumptions, not confirm them.

At one diagnostics firm, embedding validation teams from day one reduced verification iterations by 40% and accelerated regulatory clearance by three months.

When you build for testability, you build for trust, faster.

12.3.3 From Lab to Life – Real-World Readiness

Too many products pass V&V in the lab but fail in the field. Why? Because traditional V&V often ignores the real-world context.

Consider a cardiac monitoring device that passed all technical tests but struggled with adoption among elderly users due to poor app usability. Post-market validation revealed the gap. Usability-focused revalidation and UI redesign corrected it before harm occurred and improved adherence in the process.

Great V&V doesn't just confirm what works. It reveals what matters most to the patient, the clinician, and the system.

12.3.4 Risk-Based V&V – Precision over Volume

Not all features carry the same weight, yet many test strategies treat them as if they do. Smart V&V is risk-informed. It focuses depth and rigor where potential harm, complexity, or novelty is highest.

Leading teams use risk-to-test traceability matrices, where every high-severity risk is linked to at least one verification or validation activity. This sharpens focus, avoids bloated test plans, and delivers better assurance with fewer resources.

> *"We don't test everything equally. We test*
> *the most important things relentlessly."*

It is not just about proving you tested enough. It is about proving you tested what counts.

12.3.5 Automation and Digital Thread – Smarter, Not Just Faster

Digital V&V is not just a speed play, it is also a strategic enabler. Automated test scripts, model-based verification, and AI-assisted test prioritisation allow teams to test more intelligently.

One MedTech software team used AI to analyse defect trends, usage patterns, and feature risk. It re-prioritised firmware tests, accordingly, reducing test cycles by 30% without increasing escapes.

And when a requirement or risk control changes, smart systems auto-flag impacted test cases, reducing rework and preventing oversight.

This is living V&V, responsive, traceable, and always aligned.

12.3.6 V&V as a Learning Engine

Every test, pass or fail, generates insight. Yet in many organisations, this learning is lost. High-maturity teams treat V&V as part of a continuous improvement cycle, looping outcomes into design updates, requirement refinement, and future risk planning.

A system-level test failure at one company led not only to a design change but also to updates in requirement authoring and risk templates. The next cycle saw a 30% drop in test failures, not from luck, but from institutional learning.

The best teams don't just verify products. They validate learning.

> *"If you treat V&V like a checkbox, you will miss the gold it hides beneath the surface."*

12.3.7 Making V&V Visible to Leadership

To elevate V&V, it must be visible, not just as a report, but as strategic intelligence. Dashboards, governance forums, and executive reviews should surface questions like:

- What are our high-risk requirements still unvalidated?
- What usability issues were uncovered during testing?
- Which design changes were triggered by V&V feedback?

This moves the conversation from "Are we done testing?" to "Are we confident in what we have built?"

12.3.8 A Culture of Continuous Assurance

The most advanced organisations no longer treat V&V as a phase. They treat it as a mindset, integrated throughout design, development, and even post-market surveillance.

They align verification with agile sprints, use real-time dashboards to monitor readiness, and continuously re-assess priorities as products evolve.

V&V is not the end of the story. It is the discipline that shapes the story: one test, one insight, one decision at a time.

"Every failure in V&V is a flaw caught
before it becomes harm."

12.4 Documentation Reimagined: From Records to Organisational Intelligence

12.4.1 From Burden to Backbone

In many MedTech organisations, documentation is seen as an administrative necessity, something created for auditors, archived, and rarely touched again. But when treated with intent, documentation becomes far more than a regulatory artefact. It becomes the connective tissue that links product design, quality assurance, clinical insight, and organisational memory.

High-performing teams don't write documents to prove they followed the process; they write to clarify decisions, accelerate onboarding, and drive continuity. Good documentation doesn't just record what happened. It explains why it mattered and how it connects to future success.

"Documents are not just records of what we built -
they are blueprints for what we will build next."

12.4.2 Designed for Usability, Not Just Audits

Too often, documents are written for inspectors rather than the people who use them daily. This disconnect results in SOPs that are compliant on paper but ineffective in practice.

Smart organisations flip this model. They design documents for real users – operators, engineers, and regulators – with clarity, actionability, and context in mind:

- Actionable: Clear roles, responsibilities, and outcomes.
- Contextual: Embedded in real workflows and systems.
- Searchable: Easy to navigate and find critical content in seconds.

Co-creating documents with cross-functional input ensures they reflect how work actually happens, not how it is imagined.

12.4.3 From Static Files to Intelligent Systems

The future of documentation isn't in binders or static PDFs, it lives within dynamic, integrated systems that guide action in real time.

Leading MedTech organisations are investing in digital documentation platforms that:

- Embed SOPs into digital work instructions.
- Link risk controls directly to test protocols and dashboards.
- Enable real-time change tracking and traceability across the product lifecycle.

One company connected their Design History File (DHF) elements (design inputs, V&V results, risk controls) to an interactive product dashboard. This didn't just streamline audits, rather it surfaced traceability gaps proactively and accelerated cross-functional collaboration.

"Written wisdom is not overhead —
it is operational insurance."

12.4.4 Documentation as a Training Engine

Documentation is most valuable when it fuels competency development, not just recordkeeping. Organisations that treat documentation as a learning tool unlock greater compliance, confidence, and capability.

Key practices include:

- Tagging SOPs to roles and competency frameworks.
- Embedding interactive content within learning modules.
- Using annotated guides and simulations for active learning.

One firm integrated interactive SOPs into its Learning Management System. Employees navigated step-by-step scenarios with decision branches, improving both engagement and retention and reducing training time by over 20%.

12.4.5 Measuring Documentation that Delivers

Documentation should be measured not by quantity but by impact. Forward-thinking teams assess effectiveness with metrics that reflect usability, agility, and value:

- Search effectiveness: Are users finding what they need?
- Change latency: How fast do documents reflect updated processes?
- Usage rates: Which documents are being accessed and by whom?
- Training impact: How well does documentation support knowledge transfer?

These indicators elevate documentation from background noise to a strategic lever for operational excellence.

12.4.6 Final Thought – Memory that Moves the Business

In today's fast-paced MedTech landscape, documentation must evolve from archive to asset. It is not just a record of what was done. It is a reflection of how an organisation thinks, learns, and scales.

When crafted with clarity, designed for use, and connected across systems, documentation becomes more than proof of compliance, it becomes proof of competence. It supports agility, preserves institutional knowledge, and turns every decision into a building block for the next innovation.

> *"Because in a synchronised organisation, documentation doesn't slow you down. It keeps you moving in the right direction."*

12.5 Compliance as a Catalyst

When compliance is viewed as a ceiling, teams aim to "pass." But when it is treated as a foundation, they aim to excel. Throughout this chapter, we have explored how audits, verification, validation, and documentation, often relegated to the margins of innovation, can become accelerators of quality, confidence, and competitive advantage.

> *"If you want your people to value quality, show them how quality values them back."*

Audits become strategic mirrors. V&V transforms into a design ally. Documentation evolves from paperwork to knowledge infrastructure. These are not bureaucratic obligations, they are mechanisms of clarity, alignment, and assurance when embraced with purpose.

This is not just a call to reflect. It is a call to reframe. Compliance doesn't have to be a cost, rather it can be a catalyst. Ask yourself and your teams bold, practical questions:

- Are audits treated as routine inspections, or as opportunities for strategic learning and improvement?
- How often do your V&V activities inform future design decisions, rather than just validate the past?
- Is your documentation viewed as a living source of knowledge, or just a set of static records to maintain?
- Where could greater alignment between compliance, innovation, and operations unlock speed or reduce rework?
- Are your teams equipped and empowered to extract value, not just meet requirements, from core compliance activities?

As we move into the next Chapter 13: Simplified Quality: Agile Thinking, Lean Practices, & Smart QMS Tools, we take this thinking even further.

Agility isn't just about moving fast, rather it is about moving smart. In Chapter 13, we explore how MedTech organisations are stripping away unnecessary complexity and building quality systems that are lean, nimble, and resilient. We will uncover how Agile mindsets, Lean methodologies, and digital QMS tools are helping companies scale innovation without sacrificing control and how empowering people, not just processes, is the true key to transformation.

Because the future of MedTech quality isn't heavier. It is smarter.

Let us begin.

CHAPTER 13

SIMPLIFIED QUALITY – AGILE THINKING, LEAN PRACTICES, & SMART QMS TOOLS

There comes a point in every organisation's evolution when the very systems built to uphold quality begin to weigh it down. What began as structure becomes sprawl. Checklists multiply, approvals stack up, and decisions slow, not in the name of safety, but out of habit. In the pursuit of diligence, we fall into a trap: mistaking complexity for rigour.

But true quality was never meant to be a bureaucratic exercise. It was always a force for clarity, consistency, and confidence.

This chapter is a reformer's call to action, a return to first principles. Simplification is not a shortcut. It is a discipline. It asks us to build systems that are robust but elegant, compliant yet human-centric, protective without becoming obstructive. As Einstein said, *"Everything should be made as simple as possible, but not simpler."* That is the narrow path we walk.

We explore how to work smart for quality, by applying agile thinking, lean practices, and smart QMS tools to create systems that feel intuitive, not intimidating. Where doing the right thing isn't harder, it is easier. Where compliance doesn't resist innovation, it enables it.

Because true quality doesn't live in binders or bureaucratic flowcharts. It lives in clarity. It lives in trust. And increasingly, it lives in systems that think, adapt, and evolve, just like the people they serve.

In the pages ahead, we will:

- Uncover the hidden costs of complexity in quality systems
- Reimagine QMS design through Agile and Lean principles
- Leverage smart digital tools, not to replace judgment, but to elevate it
- Show how simplification, far from lowering standards, is the key to sustaining them at scale

Simplified quality isn't a trend, rather it is a survival skill. In a world that demands speed, trust, and precision, the organisations that win won't be the most thorough. They will be the most clear, nimble, and synchronised.

13.1 Why Simplify Quality?

13.1.1 Complexity: When Good Intentions Become a Burden

Most quality systems don't start bloated, but they grow to be that way. A new regulation, a past audit finding, a well-intended process fix – each adds another checklist, another approval, another layer. Over time, the system begins to bend under its own weight.

You know it has happened when a new team member asks, "Why do we need five approvals to update a form?" And no one has a good answer.

Complexity often masquerades as diligence. But instead of improving safety, it creates confusion. It delays action, obscures accountability, and makes errors more, not less, likely. In MedTech, where clarity can be lifesaving, this isn't just a nuisance. It is a risk.

"Complexity feels safe, until it
blinds you to the real risks."

13.1.2 A Turning Point: From Quality Enabler to Organisational Tax

At a fast-scaling MedTech firm, the QMS expanded from 50 to 400 SOPs in five years. Some were duplicative, many were outdated, and others were outright contradictory. Audit prep turned into a scramble. Timelines slipped. And quality was increasingly viewed not as a partner, but as a tax.

Rather than pile on more controls, the firm did something bold: they paused and launched a Quality Simplification Sprint. Every SOP, form, and approval loop was re-evaluated with three blunt questions:

- Is it necessary?
- Is it clear?
- Does it add value?

The outcome:

- SOPs reduced by 40%
- Approval loops were cut in half
- Redundant metrics were removed

The results were real:

- Product development accelerated
- Audit findings declined
- Employee trust in quality soared

Quality stopped being feared. It became understood and owned.

13.1.3 Simplicity Isn't Less, It Is Precision

Simplification doesn't mean "cutting corners." It means cutting noise.

A high-functioning QMS is not one with more documents, it is one where every document has a purpose. Where people follow procedures not out of fear, but because the process makes sense. Where compliance is natural, not nerve-wracking.

Think:

- A three-step change control process that guides, not grinds.
- A visual SOP a technician can actually follow on the floor.
- A risk matrix that lights up in a dashboard before issues escalate.

This is not minimalism. It is clarity by design.

13.1.4 Strategic Hygiene: Make Simplification a Habit

Even the best-designed QMS collects clutter over time. Approvals creep back in. Templates proliferate. Dashboards swell with stale metrics. That is why simplification isn't a one-time fix, rather it is a discipline.

World-class teams embed "QMS hygiene" into their rhythms:

- Post-audit cleanups
- Quarterly simplification reviews
- Annual documentation resets

They ask:

- Which forms go unused?
- Which sign-offs add no value?
- Which SOPs repeat themselves?
- Which metrics generate noise, not decisions?

One MedTech firm trimmed 37 SOPs to 14, halved its sign-off layers, and cut irrelevant metrics by two-thirds. Change control cycle time improved by 40%. Audit issues fell by 30%. But most importantly, employee confidence in the QMS rose sharply.

They didn't compromise quality. They removed the friction that was hiding it.

13.1.5 What Simplicity Unlocks

When simplification becomes part of the culture:

- Innovation flows more freely
- Teams take ownership
- Audits feel like reviews, not firefights
- Compliance becomes the natural output of a smart system

> *"The real test of quality isn't during an audit, it is on a Wednesday afternoon, under pressure, when the team still chooses to follow the process."*

The goal isn't less work. It is more usable work. Simplicity sustains quality at scale. Especially in an environment where both innovation and regulation are accelerating.

> *"Simplicity isn't weakness. It is maturity."*

13.1.6 Summary: The Case for Simplification

Simplification is not about doing less. It is about doing only what matters: with clarity, speed, and shared ownership.

The best MedTech organisations understand this: clarity is not the enemy of rigour. It is its foundation.

When teams can move faster, make fewer errors, and understand why each step matters, compliance becomes not just easier, but stronger.

Let simplicity be your new standard and watch quality rise.

13.2 The Agile Mindset and Lean Methods

13.2.1 From Anchors to Engines

In the early stages of MedTech growth, quality systems are often built as anchors, designed to ensure control, traceability, and compliance. But over time, those anchors can become weights. The system that once protected progress now slows it down.

This is where Agile thinking and Lean methods become transformative. Agile injects speed and feedback, while Lean strips waste and sharpens focus. Together, they turn quality from a procedural burden into a driver of responsiveness, clarity, and innovation.

We don't need heavier systems. We need smarter, lighter ones, designed to evolve with the teams they serve.

13.2.2 Living Systems, Not Monuments

Traditional QMSs often resemble monuments: rigid, formal, and hard to change. In contrast, today's leading organisations treat their QMS as a living system: adaptive, iterative, and informed by real-world use.

Agile principles provide the blueprint. Instead of overhauling the entire system, teams run short sprints, experimenting, learning, refining.

One MedTech firm revamped its CAPA review process through three rapid sprints:

- Visualised the backlog
- Eliminated duplicate reviews
- Added risk-based triage lanes

In 10 weeks: CAPA closure times dropped 35%, team participation increased, and compliance was preserved.

Agile doesn't mean unstructured, it means adaptive control. And in regulated environments, that is exactly what is needed.

13.2.3 Quality in Sprints: Progress Over Perfection

Why wait a year to revise a broken SOP when a better version could be piloted in two weeks?

A diagnostics company applied Agile sprints to improve its document control process:

- Moved approvals online
- Reduced sign-off layers
- Automated version control

In 90 days, frustration turned into flow. It wasn't perfect, but it was significantly better and improving.

Another team created a "Quality Backlog," a single-page board tracking process pain points, improvement ideas, sprint owners, and timelines. It became a living roadmap, transforming the QMS into a product under continuous development.

Progress beats perfection, always.

13.2.4 Lean Practices: Less Waste, More Value

Lean thinking doesn't mean doing more with less, it means doing what matters most.

It targets common quality system waste: excessive approvals, redundant documents, unused reports, and oversized SOPs. These inefficiencies aren't harmless: they erode trust, bury insight, and slow momentum.

One MedTech firm mapped its investigation-to-CAPA process. The delay wasn't in finding root causes, it was in formatting, review, and rework. By eliminating low-value steps and reducing approval layers, closure time dropped from 90 to 50 days.

Another reduced 60 SOPs into 25 integrated workflows, but each embedded with just-in-time guidance and risk controls.

The result:

- 30% faster quality cycles
- 40% fewer audit comments
- Stronger trust in the system

Lean isn't minimalism, it is precision.

13.2.5 Empowering the Frontlines

Agile and Lean only succeed when people, not paperwork, are at the centre.

If procedures are ignored, the problem isn't the people, it is the process. If forms are bypassed, they are likely adding friction, not value.

One organisation introduced daily quality standups across functions. No more waiting for monthly review meetings; blockers were surfaced and solved in real time.

Another appointed a Quality Product Owner. Their job: prioritise improvement ideas, own the QMS backlog, and gather user feedback. This role shifted quality from oversight to ownership: faster fixes, higher engagement, and frontline pride in the system.

> *"People don't resist quality. They resist nonsense disguised as quality."*

13.2.6 Agile + Lean = A QMS That Works

When implemented with purpose, Agile and Lean elevate, not dilute, quality. They remove unnecessary burden, amplify frontline insight, and create systems that people trust. This QMS transforms the system:

Traditional Quality → Agile + Lean Quality
Annual revision cycles → Sprint-based iteration
Bulky SOPs → Visual, usable instructions

Top-down enforcement → Compliance by design

Complex change control → Risk-prioritised fast lanes

Passive process followers → Active system co-creators

Agile fuels momentum. Lean sharpens focus. Together, they create quality systems that work for people, for regulators, and for patients.

13.2.7 Summary: Think Fast. Think Focused.

Agile and Lean aren't buzzwords. They are strategic mindsets that cut through complexity, accelerate alignment, and empower teams to own the system, not be trapped by it.

In a MedTech world where both speed and safety are non-negotiable, simplification is not a luxury. It is a necessity. Build systems that evolve. Create workflows that make sense. Give teams the clarity and the confidence to lead quality from the inside out.

When you do, compliance isn't just met. It is lived.

13.3 Smart QMS Tools and Tech Enablement

Modern MedTech organisations can't compete, nor comply, without a capable QMS. But simply digitising existing processes doesn't drive transformation. Without clarity, technology only automates confusion.

Smart QMS tools do more than digitise, they simplify, elevate, and embed quality into daily work. They don't replace judgment, they amplify it. They make doing the right thing the easiest thing.

Don't digitise chaos. Simplify first, then let technology take it further.

13.3.1 From Manual Grind to Intelligent Support

Too many quality professionals still navigate a tangle of spreadsheets, email chains, and conflicting versions. Smart QMS platforms eliminate that burden. They streamline tasks, reduce handoffs, and embed compliance into the flow of work.

A global MedTech firm cut change approval cycles from 22 to 9 days using mobile-enabled QMS software with intelligent routing. More importantly, teams shifted from chasing signatures to solving problems.

Smart tools unlocked:

- Real-time traceability
- Error reduction
- Higher engagement and accountability

Technology should remove friction, does not create new forms of it.

13.3.2 Integration – From Silos to System Thinking

A QMS cannot operate in isolation. It must be connected to PLM for design, ERP for supply chain, CRM for field feedback, and LMS for training. This integration transforms it from a document vault into a nervous system for quality.

Imagine this:
A PLM-triggered part change auto-updates the Design History File, notifies Regulatory Affairs, triggers re-training in the LMS, and prompts SOP review, all without human bottlenecks.

That is not just automation, it is proactive quality control. Integrated systems create flow, surface risk early, and ensure that changes don't fall through the cracks.

13.3.3 Maturity Over Ambition - The Tool Adoption Curve

Not every company needs AI from day one. Smart digital transformation matches tool capability with organisational maturity.

QMS Tool Maturity Model:

- Manual: Paper SOPs, Excel tracking, disconnected processes
- Digital: e-QMS for core controls like documents, CAPA, and training
- Integrated: PLM, ERP, and CRM linked to QMS for lifecycle visibility
- Intelligent: Predictive tools, AI-supported root cause analysis
- Adaptive: Systems that evolve based on user behaviour and feedback

It is not about having the most features. It is about having the right ones, at the right time.

Adoption beats ambition. Master the basics, then scale.

13.3.4 AI in Quality - From Reactive Oversight to Proactive Intelligence

Artificial Intelligence in quality management is no longer a futuristic concept, rather it is a present-day advantage. Leading MedTech organisations are already leveraging AI to shift from reactive compliance to proactive assurance, turning data overload into actionable insight.

Imagine a system that doesn't just store complaints but intelligently clusters them, flagging emerging risks before they show up in audit findings. Or a platform that scans historical CAPAs and suggests likely root causes the moment a deviation is logged. AI-powered tools can auto-draft risk assessments based on design changes, detect anomalies in audit trends, and even prioritise verification tests based on historical defect patterns.

But AI doesn't replace human judgment, it refines it. It filters noise, focuses attention, and empowers quality professionals to act faster, with greater precision. Instead of drowning in data, teams operate with clarity, able to prevent issues, not just document them.

Used wisely, AI becomes a strategic ally. Not to remove accountability, but to elevate it: by making quality smarter, faster, and more resilient than ever before.

13.3.5 Design for Humans – Not Just Compliance

Even the most powerful system is useless if it is frustrating to use. Friction is the enemy of adoption and adoption is the bedrock of compliance.

The best QMS tools:

- Match how teams actually work
- Load fast, search intuitively
- Require no training manual to navigate

Every second spent deciphering a form or struggling with a dashboard increases the chance of errors, disengagement, or workarounds.

> *"User experience isn't cosmetic, it*
> *is compliance control."*

13.3.6 Summary – Technology That Earns Its Place

Smart QMS tools aren't defined by features, they are defined by fit. The best systems:

- Simplify first, then digitise
- Connect intelligently across functions
- Evolve with maturity and scale
- Are designed for real users, not just auditors
- Automate with purpose, not to remove thinking, but to make room for it

In the future of MedTech, speed and wisdom aren't trade-offs, they are design requirements.

When systems remove friction and enable insight, compliance becomes natural. And quality becomes not just a department, but a shared discipline.

13.4 Empowering People in Quality: Training for Agile and Lean

Technology can transform systems, but only people can transform culture. A lean QMS with agile workflows is only as effective as the people who bring it to life. Empowering those people, at every level, is not a "nice to have." It is the critical multiplier that turns simplification into sustainable change.

13.4.1 Bridging the Gap Between Systems and People

Even the most advanced QMS will fail if people don't understand it, trust it, or feel ownership over it. Too often, teams are handed systems without context and asked to follow procedures they didn't shape, measured by metrics they don't believe in.

The result is disengagement. Compliance without conviction. Participation without purpose.

The real power of simplification is human. It is not in tools or templates, it is in behaviours, mindsets, and ownership. A simplified QMS must be built with people, not just for them.

13.4.2 Training Eyes to See Waste

Waste hides in plain sight unless people are trained to recognise it. That is why basic education in Lean and Agile principles is foundational, not optional. When team members are equipped to spot delays, rework, or overprocessing, small frustrations become opportunities for real improvement.

You don't need an army of Lean Six Sigma Black Belts. But you do want every team member to feel confident asking:

- Why are we doing it this way?
- Could this be simpler?
- What is the purpose of this step?

When people are trained to see waste, they start removing it without waiting to be told. That is where ownership begins.

13.4.3 From Permission to Participation

Training opens the door, but only culture determines whether people walk through it. Empowerment requires more than knowledge, it requires trust. People need to feel safe to question processes, suggest improvements, and act without fear of blame.

In many organisations, quality feels off-limits, owned by specialists, guarded by gatekeepers. But frontline teams know where the friction lives. They experience the drag of complexity every day. If you want elegant, practical solutions, ask the people who live the reality of the system.

That is why feedback mechanisms must be real, not symbolic. Move beyond suggestion boxes. Create fast, visible ways to surface ideas, test changes, and recognise contributors. When simplification becomes a shared mission, quality ownership expands exponentially.

13.4.4 The Power of the Collective Voice

One midsized MedTech company had a disengaged workforce and a bloated QMS. Employees saw quality as overhead, not opportunity. So, leadership launched a pilot program called "Lean Voice," an open platform where anyone could submit an idea to improve quality processes.

Each month, a cross-functional team reviewed submissions, prototyped solutions, and tracked impact.

In 12 months:

- 140 improvement ideas were implemented
- Training forms shrank from 10 fields to 4
- Complaint handling SOPs were rewritten based on user experience
- Satisfaction with "ease of quality processes" rose from 46% to 82%

This wasn't just operational change, it was cultural transformation. Employees didn't need more motivation. They needed trust, a voice, and a pathway to contribute.

13.4.5 Building a Culture of Contribution

Old mindsets treat quality as top-down. The new mindset sees it as co-owned. Teams don't just follow the system, they shape it.

Old mindset → Empowered culture
Quality is enforced → Quality is owned
Training is a one-time event → Learning is continuous
Simplification is risky → Simplification is celebrated
Frontline voices are ignored → Frontline insight drives innovation

When people understand the "why," are trusted with the "how," and invited into the "what is next," quality becomes more than compliance. It becomes craftsmanship.

You can't build a great system and hope people will use it.
You build great people, and they will create the system that works.

13.5 Results: A Nimble but Controlled Organisation

Simplification is not just a process change, rather it is a cultural pivot. When organisations commit to removing friction and empowering people, the benefits ripple far beyond audit readiness or document control. Speed improves. Trust deepens. And quality becomes embedded in the rhythm of daily work.

This is not theory. It is transformation you can measure.

13.5.1 The Vision Realised

A simplified QMS does more than streamline documentation or accelerate workflows, it changes how teams think, decide, and collaborate. What once felt bureaucratic becomes intuitive. What once slowed progress becomes a catalyst for it.

In high-performing organisations:

- Quality systems adapt with agility
- Accountability becomes natural, not forced
- Teams no longer ask for permission to act, they simply act, because the system enables it

You know the system works when people say:

> *"We didn't simplify for auditors. We simplified for us. The auditors just appreciated it."*

13.5.2 Compliance by Design

One of the most powerful benefits of simplification is that compliance becomes a natural outcome, not a scramble or a stress point.

Audit prep stops being a fire drill. Teams walk into inspections calm, confident, and prepared, because the way they work is the way they document.

After embedding the 'compliance by design' approach at a MedTech startup, the QA leader reflected to me with pride:

> *"I stopped worrying about what the SOP said. Because I knew the way we worked was the SOP."*

That is compliance by design. Where practice and policy are aligned. Where clarity removes the need for reminders, and where systems enforce quality simply by being well-structured and usable.

13.5.3 Speed, Clarity, and Confidence

Simplification doesn't just tidy the house, it accelerates everything inside it.

Real-world results from MedTech organisations that embraced lean, agile, and smart QMS design include:

- Change control cycle time reduced from 40 to 15 days
- CAPA closures 40% faster after streamlining workflows
- Internal audit findings down by 50%
- Stronger cross-functional alignment between Quality, RA, and Engineering
- Thirty-five percent improvement in employee confidence in QMS usability

Quality became easier to follow, easier to own, and harder to ignore. Dashboards replaced binders. Ownership replaced oversight. And compliance stopped feeling like a chase, it became a daily habit.

When simplicity becomes the system, excellence becomes the habit.

13.5.4 A Story of Cultural Rebirth

At one mid-sized MedTech start-up company, complexity had quietly eroded morale and performance. Long approval chains, slow decisions, and tense audits were the norm.

Leadership chose not to patch the problem, but to reset it.

A system-wide simplification initiative was launched with full transparency and team engagement.

Key changes included:

- Eliminating 120 obsolete forms
- Consolidating 37 SOPs into 15 lean workflows
- Deploying a smart e-QMS with real-time dashboards and mobile access
- Training every employee in Lean and Agile principles

Within 12-18 months, the transformation was unmistakable:

Metric	Before	After	Improvement
Document cycle time	32 days	18 days	↓ 45%
Internal audit findings	12	6	↓ 50%
Employee QMS satisfaction	52%	87%	↑ +35 pts
CAPA closure time	65 days	39 days	↓ 40%

But the true shift was deeper than metrics. Employees started caring again. They raised issues earlier. Shared solutions faster. Suggested improvements without being asked.

"Simplification didn't dilute the system, rather it made it human. And that made all the difference."

13.6 Work Smart for Quality

13.6.1 Simplicity as the New Standard

In a world that rewards speed, precision, and adaptability, complexity is no longer a sign of maturity, it is a sign of inertia. World-class organisations are learning that true rigour comes not from doing more, but from doing what matters, with clarity and purpose.

The best quality systems don't impress by their volume. They inspire through their usability. They aren't tolerated, they are trusted. They don't slow teams down, rather they remove friction so teams can fly.

Simplicity is not the opposite of control. It is the enabler of it. When systems are clean, focused, and human-centred, excellence becomes the path of least resistance.

13.6.2 A New Lens for Tomorrow

Every form, workflow, and SOP is an opportunity to choose better. Not lazier – better. With clearer intent, sharper focus, and smarter design.

As you return to your own systems, ask yourself:

- If we rebuilt this from scratch, would we build it this way?
- Does this step serve a purpose or just a habit?
- What is the worst that could happen if we simplified?
- What is the best that could happen if we did?

These are not shortcuts. They are signals of leadership. They show the willingness to evolve, and the courage to do less, better.

13.6.3 Flexibility and Focus – The Heart of SHIFT

In the SHIFT model, the "F" stands for Flexibility, but also for Focus. Focus is what ensures that simplicity doesn't become sloppiness. It channels flexibility into precision and speed into substance.

When your QMS is lean, teams move faster.
When it is smart, compliance becomes invisible.
When it is trusted, quality becomes reflex, not obligation.

Simplicity isn't about lowering the bar. It is about raising the floor, so every team operates from a place of clarity, capability, and confidence.

The future will not reward those who do more, it will reward those who do less, brilliantly.

13.6.4 A Foundation for the Future

This chapter has not been about doing less. It has been about doing better, with less drag, more purpose, and systems that are built to empower.

When simplification is pursued with intention:

- Time and energy are reclaimed for value-creating work
- Risk is managed by design, not by layering control
- Compliance becomes embedded, not bolted-on
- Trust grows, across teams, with regulators, and with customers
- Organisations gain resilience to flex, scale, and adapt to the future

Simplicity isn't a soft option. It is a strategic edge. It makes room for innovation. It reduces burnout. And it sends a message: we care enough to make it clear.

13.6.5 Start Now

If there is one takeaway, let it be this:

Start small. Simplify one form. Shorten one approval chain. Rebuild one process with the end-user in mind. Then watch the momentum build.

Because excellence doesn't begin with complexity. It begins with clarity and one cleaner step.

Start now. You will never look back.

This is not just a call to reflect. It is a call to simplify. Complexity doesn't always mean control, sometimes, it is the barrier. Ask yourself and your teams bold, practical questions:

- Where has your quality system become more complex than it needs to be and why?
- Are your teams equipped with lean tools and agile methods that streamline work and improve flow?
- How often do processes get designed around audits or legacy habits, rather than value and usability?
- Are your digital QMS tools enabling efficiency and insight or adding new layers of friction?
- What is one area of quality where simplification could create immediate clarity, speed, or engagement?

As we close this chapter on simplification, we turn to an area where clarity is not just empowering, it is essential.

Chapter 14 explores the digital backbone of modern MedTech, where devices are connected, data flows continuously, and risk never sleeps. In this world, quality isn't just about what is documented, it is about what is defended, detected, and designed for resilience. From cybersecurity and interoperability to real-time safety systems, we dive into the disciplines that safeguard trust in a hyper-connected environment.

Because in the end, it is not just about building great products. It is about keeping them secure, connected, and worthy of the confidence patients and professionals place in them: every second, without exception.

THE DIGITAL LEAP - FROM ALGORITHMS TO ACTION

We now enter the next frontier, not of theory, but of transformation.

Digital disruption is no longer coming to MedTech. It is here. And it is rewriting the rules for how we design, deliver, validate, and sustain healthcare technologies. The organisations that will lead, not follow, are the ones who embrace this leap with clarity, caution, and coordination.

This section explores what it truly means to be a synchronised, digitally fluent MedTech organisation in a world shaped by AI, real-time data, and post-market transparency. It is not about chasing trends. It is about anchoring innovation in trust, strategy, and systems that scale responsibly.

We begin with Chapter 14: Secure, Connected, Reliable – Cybersecurity, Interoperability, & Real-Time Safety, because in an increasingly connected world, safety is no longer confined to the product, it extends into the cloud, the hospital network, the app on a patient's phone. This chapter unpacks the modern risk lifecycle: continuous, distributed, and dynamic. Cybersecurity becomes more than IT's job, it becomes a product requirement, a regulatory expectation, and a brand differentiator. Interoperability is treated not as a technical checkbox but as the foundation of safe, smart conversations between systems. And real-time safety isn't a dream, rather it is a design goal.

From there, we turn to the transformative power of artificial intelligence and simulation in Chapter 15: AI, Digital Twins, & the Next MedTech Leap. AI and digital twins are more than tools, they are ways of thinking. Whether it is a model predicting patient deterioration or a virtual prototype simulating design performance, these technologies can radically compress timelines, improve precision, and elevate safety, if embedded responsibly. This chapter explores both the promise and the pitfalls, showing how synchronised organisations integrate AI not just into products, but into their culture, QMS, and validation pipelines.

But innovation doesn't stop at launch. In Chapter 16: From Data to Decisions – RWE, Virtual Trials, & Post-Market Maturity, we explore how synchronised organisations treat post-market as a strategic advantage, not a regulatory obligation. Real-world evidence (RWE) becomes a living feedback loop, not just proving effectiveness but improving it. Virtual and hybrid trials extend access and speed. Surveillance evolves into insight. The result? A learning organisation that adapts at the pace of reality.

Finally, Chapter 17: The Submission Shift – Turning Compliance into Competitive Advantage brings us full circle. In a digital, globalised ecosystem, regulatory documentation is no longer a final hurdle, it is a strategic narrative. This chapter explores how forward-looking companies craft submissions that are clear, scalable, globally aligned, and built from the start, not stitched together at the end. With smart tools, international foresight, and cross-functional authorship, the dossier becomes a reflection of synchronisation in motion.

Together, these four chapters push the reader into the future, grounded in reality, but fuelled by readiness. This is the digital leap: the moment when MedTech stops reacting to change and starts orchestrating it.

And the synchronised organisation? It is not overwhelmed. It is already in stride.

CHAPTER 14

SECURE, CONNECTED, RELIABLE – CYBERSECURITY, INTEROPERABILITY, & REAL-TIME SAFETY

The evolution of MedTech has moved from devices that operate in isolation to systems that live in real-time networks: sensing, communicating, and adapting. A pacemaker no longer just paces, it transmits telemetry to cloud platforms. An infusion pump doesn't just deliver medication, it reads from EMRs and adjusts flow based on algorithms. Today, the clinical frontier is not only physical, but also digital.

But with this digital power comes digital responsibility. Connectivity doesn't just enable care, it can also introduce vulnerabilities. A misconfigured API, a delayed software update, or an unsecured signal can become a safety event. In this new era, quality is judged not only by design precision but by a product's cybersecurity posture, interoperability fluency, and real-time resilience.

This chapter explores what it means to deliver safe, connected, and intelligent MedTech, where cybersecurity is foundational, not optional, where systems speak clearly to one another, and where devices protect life even when things go wrong.

We will move beyond fear-based security thinking and toward proactive trust-building through integrated governance, intelligent monitoring, and human-centred system design. Because in a connected world, a device's function is only as good as its ability to protect, adapt, and communicate with confidence.

The future belongs to MedTech leaders who aren't just innovative but vigilant. Not just compliant but trusted. Let us explore how to design systems that are secure by default, connected with clarity, and reliable under pressure. Because in this next chapter of care, trust isn't a byproduct, it is the product.

14.1 The Modern Risk Lifecycle: Connected, Continuous, Critical

The FDA issued a public warning: certain insulin pumps could be tampered with remotely. A flaw in their wireless protocols meant someone within range could alter dosage, potentially with fatal consequences. These weren't hypothetical threats, they were real devices, in real patients, with life-threatening vulnerabilities.

Welcome to the age of connected MedTech, where clinical efficacy and cybersecurity are no longer separate concepts.

14.1.1 Expanding the Boundaries of Quality

Modern devices no longer operate in isolation. A pacemaker isn't just a physical implant, it is a digital node in a complex network of smartphones, cloud servers, and clinical systems. A wearable doesn't just record data, it transmits, syncs, and responds. This transformation expands the quality landscape beyond sterility, calibration, and mechanical integrity. Now, quality must account for digital risks:

- Device hijacking through unsecured interfaces
- Malicious code injected during software updates
- Cloud service outages that compromise real-time functionality
- API mismatches leading to failed alerts or misrouted data

A product's safety is now defined not just by whether it works, but whether it works securely, reliably, and intelligently across an interconnected ecosystem.

14.1.2 From Reactive to Real-Time Risk Management

Traditional risk management has been retrospective, triggered by audits, complaints, or CAPAs. But in a connected world, this lag is unacceptable. Risk now evolves minute by minute. Each patch, integration, or data exchange introduces new potential failure modes.

Leading organisations adopt a real-time risk posture:

- Detect digital anomalies as they occur
- Monitor connected assets continuously
- Deploy secure patches with confidence
- Run cyberattack simulations as part of regular validation
- Incorporate live threat intelligence into QMS dashboards

This is a fundamental mindset shift, from "Have we tested enough?" to "Are we prepared for what is next?"

> *"In a connected world, you are not just building*
> *a device — you are building trust across every*
> *byte, every signal, every heartbeat."*

14.1.3 A Maturity Model for Connected Risk

Organisations move through four levels of digital risk maturity:

Level	Mindset	Risk Approach	System Behaviour
1	Static Compliance	Reactive fixes post-failure	Audit-driven, siloed, slow to adapt
2	Process-Oriented	Periodic risk reviews	Improved control, limited flexibility
3	Adaptive & Integrated	Proactive mitigation planning	Cross-functional, iterative improvements
4	Predictive & Real-Time	Continuous monitoring + AI-enabled	Risk signals actively guide operations

Level 4 is the target state, where quality systems don't just record risks, they respond to them in real time.

14.1.4 From Theory to Action

To transition from concept to execution, ask yourself:

- How many of your devices rely on continuous connectivity?
- Can your QMS respond to a live cyber threat within hours?
- Are patches treated like emergency fixes or planned design changes?

Start embedding connected risk practices now:

- Add digital threat modelling to design reviews

- Include interoperability testing in the Design Master Record (DMR)
- Conduct Connected Device FMEAs, mapping failure across software, network, and cloud interactions

Some leaders run monthly "Digital Risk Sprints," brief, focused cycles where cross-functional teams proactively assess known Common Vulnerabilities and Exposures (CVEs), test recovery protocols, and deploy updates before they become urgent.

14.1.5 The Living System Mindset

Thriving in this environment requires more than tools, it demands a living-system mindset.

That means systems that adapt, architectures that evolve, and teams that view safety as a moving target. Modern MedTech companies now:

- Integrate threat intelligence into product roadmaps
- Adjust design specifications in response to digital risks
- Incorporate security metrics into quality KPIs

If your system only protects patients the day it was certified, it is not a resilient system, it is a frozen snapshot. In the connected era, quality isn't static. It is demonstrated daily, across every heartbeat, every signal, and every update.

> *"Trust is no longer earned once, instead*
> *it must be re-earned continuously."*

14.2 Cybersecurity and Governance: From Protection to Trust

A vulnerability was discovered in a cloud-enabled wearable cardiac device. Within five days, the manufacturer verified, mitigated, and patched the issue, before public disclosure. Users were informed, regulators updated, and trust, far from eroding, was strengthened.

This is the reality of connected MedTech today: if it isn't secure, it isn't safe. And if it isn't safe, it cannot be trusted.

Medical devices now function as digital endpoints, exchanging data, syncing with cloud services, and integrating into hospital infrastructure. In this environment, cybersecurity is not a bolt-on IT function. It is foundational to product integrity and patient safety.

14.2.1 Security by Design, Not by Patchwork

Security must begin on day one. Retroactive fixes, post-market hardening, or waiting for threats to emerge are not viable strategies. A secure product is intentionally architected, not reactively defended.

Secure-by-design principles include:

- Strong authentication and role-based access controls
- End-to-end encryption from device to cloud
- Secure boot sequences that verify software before execution
- Fail-safe update mechanisms that avoid service disruption

Beyond architecture, teams must proactively assess threats. Threat modelling, attack surface analysis, and code hardening are now as critical as verification testing and biocompatibility.

"Designing without cybersecurity is like
building a hospital without doors. Everything
may work – until it is exposed."

14.2.2 Regulatory Alignment – Global Cyber Norms

Regulatory bodies now treat cybersecurity as a core product requirement. Across jurisdictions, expectations are shifting from best-practice to baseline:

- The FDA requires a Software Bill of Materials (SBOM), vulnerability management plans, and coordinated disclosure protocols.
- EU MDR (Annex I) demands protection against unauthorized access and system compromise.
- ISO/IEC 81001-5-1 and IMDRF guidance emphasise lifecycle-wide cybersecurity planning with traceability, threat assessment, and post-market vigilance.

A modern QMS must reflect these realities. Security artifacts, such as threat models, risk assessments, update procedures, and penetration test results, should be integrated into the design history file and technical documentation.

Compliance now depends not only on what you build, but how you protect it over time.

14.2.3 Agile Response – Speed, Clarity, Containment

No system is invulnerable. What separates resilient organisations is their speed of response and clarity of communication when an incident occurs.

Top-performing MedTech firms:

- Conduct penetration testing before release and after major updates
- Maintain clear, well-drilled incident response protocols
- Publish responsible disclosure policies that invite ethical feedback
- Patch critical vulnerabilities within five days of detection

Response time is no longer a technical issue, rather it is a business risk and a reputational differentiator. A fast, transparent, coordinated response can turn a breach into a moment of earned trust.

14.2.4 Cross-Functional Ownership - Security is Everyone's Job

Security cannot be delegated to IT alone. In connected MedTech, it is a shared responsibility:

- R&D must integrate secure coding and testing into the development lifecycle
- Regulatory Affairs must align cybersecurity documentation with submission expectations
- Post-market Surveillance must monitor field threats with the same diligence as product complaints
- Clinical and Safety Teams must evaluate downstream impacts of potential software failures

True cyber maturity requires both process and culture. Everyone, from engineer to executive, owns the promise of protection.

"You don't build trust by hiding breaches. You build it by being the first to act on them."

14.2.5 Clinical Environment Collaboration – Extending the Security Perimeter

A device's security must hold in the complexity of real-world clinical environments. It must operate within:

- Hospital networks burdened with legacy systems
- Variable EHR platforms with differing levels of integration
- Clinical workflows where uptime is mission-critical and interoperability is non-negotiable

Forward-looking manufacturers:

- Involve hospital IT teams early in development
- Share APIs and interface specs with integration partners
- Simulate real-world network conditions during validation
- Run cross-vendor integration tests before market deployment

Security doesn't end at your firewall. It extends into the hospital, the cloud, and ultimately, into the patient experience.

14.2.6 Governance Structures – Building Digital Risk Discipline

Many leading organisations are now forming Digital Risk Boards, a cross-functional governance bodies responsible for cybersecurity oversight, threat intelligence, and policy alignment.

These boards typically include:

- Quality and Regulatory leaders
- Product Security Engineers
- Design and Cloud Operations leads
- Legal, IT, and third-party security advisors

Meeting quarterly (or more frequently), the board:

- Tracks priority CVEs and active threat vectors
- Aligns product updates with emerging vulnerabilities
- Reviews incident readiness and patch closure metrics
- Coordinates regulatory messaging and customer communications

This is not window dressing, it is a core mechanism for aligning product, risk, and trust in the digital era.

14.2.7 Trust as the Product

Today, a secure device is not a bonus, it is the expectation. And trust is not earned once at approval, it must be reaffirmed with every update, every alert, every connection.

To build digital trust, MedTech devices must be:

- Secure by design
- Proactive in detection
- Fast in response
- Transparent in communication
- Supported across the full lifecycle

Security isn't a technical checkbox. It is a strategic differentiator and a promise to patients, clinicians, and regulators alike.

"You don't protect people from technology,
rather you protect them through it."

14.3 Interoperability: Designing Safe Conversations Between Systems

The era of stand-alone medical devices is over. Today's devices must communicate across platforms, systems, and settings. A wearable glucose monitors syncing with a smartphone, an infusion pump interfacing with an EHR, a ventilator alerting a central dashboard – these aren't exceptions. They are expectations.

Interoperability has become as critical to safety as sterility or software validation. If a device can't "speak" clearly and reliably, it risks being misunderstood or ignored by the systems meant to protect patients. In connected care, integration is not just convenience. It is a matter of clinical precision and operational trust.

14.3.1 The New Language of Healthcare

Interoperability is the ability of different systems and devices to exchange, interpret, and act upon shared data. But it is not enough to connect. Devices must communicate meaningfully, safely, and without ambiguity.

For clinicians, this means unified, real-time patient views, device data merged with labs, medications, and histories. For patients, it means seamless handoffs between home, hospital, and mobile care. For systems, it demands fluency in shared standards such as HL7 FHIR, DICOM, IHE profiles, and Bluetooth Health protocols.

Regulators increasingly expect conformance with these frameworks, not as technical preferences, but as proxies for readiness, predictability, and patient safety in a data-driven world.

"Innovation speaks loudly, but
integration whispers trust."

14.3.2 Simulation Over Assumption – Validating in the Real World

A device that connects in a lab may fail in a hospital. Integration isn't just about handshake success, it is about sustained, reliable operation in the real world.

Traditional testing often falls short, relying on static environments or one-way message checks. Modern interoperability validation requires:

- System-level simulations with production EHRs
- Multi-vendor connectathons under load conditions
- Scenario testing for timestamp mismatches, alarm propagation, and unit conversion errors

In one instance, a device passed all integration tests but failed in deployment due to misaligned log timestamps, undermining traceability during an adverse event review. It wasn't a hardware issue. It was a communication breakdown.

Real-world simulation must become standard, not an afterthought.

"It's not enough to say "we connect." You
must prove we connect safely."

14.3.3 Designing for Clarity – Eliminate Ambiguity at the Interface

The interface is where clinical systems meet. It must be bulletproof.

Even small discrepancies – wrong units, message delays, ID mismatches – can lead to catastrophic results. A smart pump interpreting dosage in "mg" instead of "mL." An alert that doesn't reach the nurse station. A duplicated patient ID across systems.

To prevent these risks, development teams must:

- Create precise, version-controlled interface specifications
- Validate across multiple environments, vendors, and clinical workflows
- Embed interoperability as a design input and validation checkpoint
- Treat data fidelity as part of safety, just like dosage accuracy or labelling clarity

This is not optional. EU MDR's Annex I requires that connected devices do not negatively impact each other's performance. The FDA's guidance expects robust documentation on intended uses, limitations, and risk mitigations across integrations.

> *"In connected healthcare, integration errors are not technical bugs, they are quality failures."*

14.3.4 The Interoperability Dividend – Why It Matters

Done right, interoperability delivers measurable value:

- Reduced manual transcription and data entry
- Faster alarm routing to the right caregiver, at the right time
- Seamless device data archiving into EHRs and clinical decision systems

- Lower clinician burden and increased adoption
- Faster hospital procurement approval through demonstrated ecosystem compatibility

In one case, a device manufacturer uncovered a critical medication unit mismatch during EHR integration testing, preventing a potential dosing error before market release. As a result, hospital onboarding time dropped, nurse confidence rose, and the device was rapidly adopted across facilities.

Interoperability doesn't just reduce risk, it accelerates trust.

14.3.5 From Technical Feature to Strategic Differentiator

In today's procurement landscape, a device that doesn't integrate is a device that won't be selected. Hospitals expect seamless operation within complex digital ecosystems. That means:

- Integration-ready APIs
- Clear documentation
- Proven testing history in clinical environments
- Willingness to collaborate on validation and deployment

Interoperability is now a competitive advantage. Not just because it enables care, but because it reduces risk, amplifies safety, and strengthens the trust clinicians place in your product.

The devices that win the future won't just be clinically excellent. They will be digitally fluent, designed to communicate, collaborate, and contribute safely to the moments that matter most.

14.4 Real-Time Monitoring and Fail-Safe Systems

Connected medical devices are no longer passive instruments, they are dynamic, intelligent agents embedded in complex care ecosystems. Where once safety depended solely on rigorous premarket validation, today it also depends on continuous performance monitoring, live feedback, and system resilience under pressure.

Real-time monitoring and fail-safe design are not future aspirations. They are the new imperatives of modern MedTech. Because in today's connected environment, safety doesn't end at product release, it begins there.

"If you can't monitor it, you can't protect it."

14.4.1 From Passive Tools to Active Guardians

Modern medical devices must do more than function, they must observe, respond, and protect. Embedded diagnostics, connectivity, and telemetry enable devices to:

- Detect subtle deviations in performance or patient response
- Alert clinicians to emerging issues before symptoms appear
- Trigger remote updates or maintenance cycles without delay
- Learn from field use to improve reliability over time

In one instance, a wearable ECG detected firmware anomalies affecting a specific batch. Within 24 hours, telemetry flagged the issue, triage was completed, and a targeted patch was deployed, averting a potential wave of sync issues and reinforcing user confidence.

Real-time monitoring transforms devices into sentinels: always watching, always ready to act.

> *"A safety net is only strong if it catches what*
> *matters — and lets the rest pass."*

14.4.2 Prioritised Alerts – Insight Over Noise

Not all data is created equal. Poorly configured systems overwhelm users with false alarms, creating alert fatigue and disengagement. Effective real-time systems:

- Prioritise high-risk conditions
- Suppress redundant signals
- Deliver alerts in context, aligned to patient state, workflow, and caregiver availability

A safety net is only effective if it catches what matters. Quality is not just measured by detection, but by meaningful, actionable insight.

14.4.3 Fail-Safe Thinking – Preparing for Disruption

Connectivity is a double-edged sword. It enables powerful new capabilities, but introduces new dependencies: on networks, cloud infrastructure, and firmware synchronization.

Fail-safe design ensures that when (not if) disruptions occur, the system remains safe. Devices must be engineered to:

- Revert to safe, validated local modes if cloud access is lost
- Buffer data for delayed transmission and resynchronisation
- Maintain safe operations across multiple failure scenarios, including power loss, latency spikes, or server downtime

Dual-path communication, autonomous fallback modes, and secure local overrides aren't luxuries, they are table stakes for trust in high-acuity environments.

If your system fails safely, you have protected the patient, even when the network fails you.

"Connectivity is a feature. Safety is a promise."

14.4.4 Real-Time Data as a Living Quality Asset

Real-time data is no longer just operational telemetry. It is a continuous source of insight for quality improvement and risk reduction. Leading companies now treat connected field data as a live CAPA input, enabling:

- Early detection of systemic issues
- Predictive maintenance planning
- Field-based failure trend analysis
- Continuous improvement cycles based on actual use

One company identified a subtle sensor drift pattern in a ventilator via cloud-streamed data. It traced the anomaly to a manufacturing deviation in a specific supplier batch. A proactive maintenance campaign prevented any adverse events, strengthened supplier oversight, and earned regulatory commendation for pre-emptive vigilance.

Real-time data shortens the distance between issue and intervention, from weeks to hours.

14.4.5 Redefining Quality for the Connected Era

Traditional quality models anchored around static milestones: design freeze, submission, release. But in connected MedTech, the product is never "done." Devices evolve through updates, environmental learning, and live feedback.

Modern quality is:

- Continuous: updated with each patch and integration
- Dynamic: adapting to real-world use and user behaviour
- Collaborative: shaped by feedback from the field and the cloud

This demands QMS systems that support version control, cloud-based change management, and ongoing performance validation. It also requires teams that are trained not just to build, but to listen, learn, and iterate, long after launch.

Real-time monitoring and fail-safe design are not just technical capabilities. They are the new definition of excellence, where safety, insight, and resilience co-exist by design, not coincidence.

"Today's device isn't just a tool. It is a teammate —
watching, learning, and protecting in real time."

14.5 Trust: The Ultimate Product

In the analogue era, quality was measured by what could be seen, touched, and mechanically verified. Precision in tolerances, clarity in labelling, sterility in packaging. But in the connected era, quality is often invisible, encoded in encrypted transmissions, embedded in secure firmware, and expressed through milliseconds of uptime.

Trust is no longer a passive outcome. It is the product. And in connected MedTech, that trust is continuously earned, across every signal, update, and interaction.

14.5.1 From Secure Design to Sustained Confidence

A secure design at launch is only the beginning. What matters more is how the system evolves:

- Can it receive validated security patches without disrupting care?
- Does it adapt to changes in hospital IT infrastructure?
- Is it resilient to emerging threats, both clinical and digital?

A modern medical device must be judged not just by what it does, but how it behaves, when it connects, when it fails, when it updates, and when it protects. Every one of these moments either reinforces or erodes confidence.

14.5.2 The New Quality Triad: Secure, Connected, Reliable

To be considered high-quality in the connected age, a device must now meet a new triad:

- Secure by Default: Designed with threat modelling, authenticated access, encrypted data, and proactive cyber hygiene.
- Connected by Design: Interoperable across platforms, readable by systems, and aligned with standards like HL7 FHIR, DICOM, and IHE.

- Reliable in Real Time: Monitored continuously, capable of alerting meaningfully, and designed to fail safely without compromising patient outcomes.

These pillars are not engineering afterthoughts. They are central to how clinicians adopt, patients trust, and regulators approve modern MedTech systems.

14.5.3 Building Trust Through Every Interaction

Trust is not won with certificates or audits. It is earned:

- In how quickly a vulnerability is disclosed and resolved.
- In how accurately a device integrates with a hospital system.
- In how reliably a device performs during a crisis.

It is embedded in quiet moments, when a system correctly routes an alert, syncs patient data without delay, or prevents harm by flagging degradation before failure.

The most innovative teams build this trust through transparency, preparedness, and intent. They treat cybersecurity, interoperability, and monitoring not as burdens, but as extensions of care.

14.5.4 From Product Release to Product Readiness

Launching a product is no longer the finish line. It is the starting point of an ongoing relationship between device, patient, provider, and environment.

Quality now lives in how systems are maintained, how updates are verified, and how anomalies are resolved before escalation.

Post-market surveillance has evolved into post-market performance, and the best organisations are embracing it as a core discipline of digital maturity.

The systems that win in the future won't just be smarter, they will also be more trusted. Because in healthcare, intelligence without integrity is a risk. Innovation without vigilance is a liability.

14.5.5 Practical Next Step – Stress Test for Trust

If you are looking to put this mindset into practice, start here:

- Choose one connected workflow, such as device onboarding, firmware update, or clinical data exchange
- Map where cybersecurity, interoperability, and reliability intersect
- Identify the weakest point. Ask: "If this fails, how fast do we know? What is our next move?"

This small exercise will build operational reflexes for the larger digital transformation ahead.

14.5.6 Closing Thought – Trust is a Living Cycle

Trust is not built once and stored away. It must be renewed continuously:

- Earned in design
- Proven in the field
- Maintained through transparency
- Strengthened with every secure connection

In the connected world, trust is a verb. And in MedTech, it may be the most vital action your product performs.

Because patients don't just rely on your technology. They rely on your vigilance.

14.5.7 Looking Ahead: The Next Leap

This is not just a call to reflect. It is a call to protect. In a connected world, safety, security, and trust must be designed in, not added later. Ask yourself and your teams bold, practical questions:

- How well does your current risk lifecycle account for the continuous, real-time nature of connected devices?
- Is cybersecurity embedded into your product development and governance, or treated as a final-stage review?
- Are your systems truly interoperable in a way that protects data, ensures performance, and enables safe integration?
- How equipped is your organisation to detect, respond to, and learn from safety signals in real time?
- What would it take to make digital trust, not just product performance, your ultimate competitive advantage?

As we close this chapter on secure, connected, and real-time safety, we shift our gaze forward, to the horizon of what is next. In next chapter, we explore the transformative potential of AI and digital twins in MedTech. Not as science fiction, but as science applied with responsibility.

"If your device is smart enough to connect,
it must be smart enough to protect."

We will examine how machine learning is accelerating diagnostics, how digital twins are reshaping prototyping, and how virtual trials are redefining clinical evidence. But we will also explore the ethical guardrails, regulatory considerations, and quality system adaptations required to make this leap responsibly.

Because the future of MedTech is not just about smarter devices, it is about synchronising intelligence, safety, and trust at scale.

Let us now step into that future.

CHAPTER 15

AI, DIGITAL TWINS, & THE NEXT MEDTECH LEAP

T he future is no longer approaching, it is already unfolding around us. In MedTech, what once lived in science fiction is becoming operational science. Artificial intelligence (AI) is augmenting human insight, helping clinicians detect anomalies invisible to the eye. Digital twins (dynamic, data-driven virtual replicas of devices, organs, and even patients) are enabling us to simulate complex scenarios that would take years of physical testing to observe.

Step into tomorrow's operating room. A surgeon sees more than anatomy, rather they see predictive overlays powered by AI. A digital twin of the patient has already rehearsed the procedure dozens of times, helping anticipate complications. These are not distant visions. In leading centres, this is happening now.

But with exponential power comes exponential responsibility. We are entering an era of predictive, adaptive, and highly personalised healthcare, one that demands new frameworks for safety, ethics, and evidence. It is no longer enough to build devices that work. We must ensure they learn safely, evolve transparently, and protect by design.

The synchronisation journey we have mapped so far, across people, systems, and quality mindsets, has laid the foundation. This chapter builds on that by exploring how AI, digital twins, and virtual trials are reshaping MedTech innovation. More importantly, we will examine how to lead this leap responsibly, embedding trust, ethics, and rigour into every algorithm, model, and decision.

This isn't just the next technological transformation. It is the next transformation of trust, and it starts now.

15.1 A Vision of AI in MedTech: Leaping Into the Future

15.1.1 From Science Fiction to Surgical Reality

The future of healthcare isn't arriving, it is integrating. AI-powered diagnostics, predictive algorithms, and dynamic digital replicas are no longer experimental novelties. They are active components in operating rooms, ICUs, and even patient homes.

Imagine a surgical suite where clinicians wear augmented reality glasses, layered with predictive insights: real-time anatomical overlays, patient-specific risk forecasts, and intelligent suggestions based on millions of prior cases. Meanwhile, a digital twin of the patient, fed by imaging, vitals, and EHR data, has simulated the procedure dozens of times in advance, flagging possible complications. This is not science fiction. It is the foundation of the next leap in care delivery.

15.1.2 AI as an Extension of Human Judgment

In MedTech, AI is most powerful not when it replaces human decision-making, but when it augments it. Trained on vast datasets from wearables, imaging, genomics, and real-world evidence, AI can surface insights faster, more accurately, and often earlier than traditional methods.

AI becomes a clinician's second set of eyes: tireless, pattern-aware, and continuously learning. It doesn't make decisions for us. It improves the quality and speed of the decisions we make. When implemented responsibly, AI shifts healthcare from reactive response to proactive intervention.

> *"The best AI doesn't take decisions away from clinicians — it gives them better ones to make."*

15.1.3 Beyond Automation – Toward Machine Reasoning

The leap from automation to intelligence is profound. Today's systems don't just follow instructions, they generate hypotheses, reason through possibilities, and support complex clinical trade-offs.

Advanced AI is now capable of asking: Why did this trend shift? What intervention is most likely to succeed based on this patient's full context? How might this therapy interact with co-morbid conditions or medications? This is not just about efficiency. It is about empowering teams to care more precisely, more safely, and more individually than ever before.

15.1.4 Digital Twins – Simulation as Strategy

A digital twin is more than a model. It is a real-time, high-fidelity virtual replica of a device, organ, or patient, built from sensor inputs, physiological data, imaging, and historical records. Unlike static simulations, digital twins continuously evolve, responding to new inputs and enabling scenario planning at clinical scale.

Already, digital hearts simulate arrhythmias to test intervention strategies. Orthopaedic companies use digital limbs to test implant durability under real-world gait patterns. ICU teams monitor high-risk patients through predictive twin models that simulate deterioration trajectories hours or days before symptoms appear.

Digital twins are not just tools for design. They are becoming strategic partners in care, regulation, and risk management.

15.1.5 Redefining Innovation and Safety

Together, AI and digital twins are redefining what it means to innovate responsibly. The new benchmark is not just performance, but adaptability. A successful product is not only accurate at launch, it must also remain accurate as conditions, data, and patient contexts evolve.

This demands a shift from static V&V to living, learning validation systems. Algorithms must be monitored like post-market products. Twins must be governed like clinical assets. Innovation must be designed to learn safely and to explain how it learns.

15.1.6 The Responsibility Leap

With intelligent systems comes deeper accountability. AI models can drift. Twins can become outdated if data inputs are biased or incomplete. Left unchecked, these technologies can erode trust or amplify disparities.

That is why the leap ahead is not only technical, but also ethical. It requires embedding explainability, auditability, and inclusivity into every layer of development. Transparency is no longer optional. It is what earns adoption, from regulators, clinicians, and patients alike.

"Trust is not built by hype. It is earned by outcomes."

15.1.7 Reframing the Future of MedTech

This is the most pivotal moment in MedTech since the first device was implanted. The question is not whether we adopt AI or digital twins. The real challenge is whether we do so with discipline, foresight, and humanity.

The goal is not just smarter devices. It is more synchronised systems. Not just code that learns, but cultures that do.

As we continue, we will explore how AI and digital twins are already reshaping diagnostics, device design, clinical validation, and surveillance. But before diving deeper into those use cases, let us examine where this transformation is already delivering value and how we can harness it responsibly and safely.

15.2 AI in Action: Applications and Opportunities

"The future is already here – it is
just not evenly distributed."
– William Gibson

AI is no longer an experimental add-on in MedTech, it is becoming foundational. From diagnostic precision to operational intelligence, AI is transforming how companies develop, deploy, and improve devices. But success depends not on ambition alone, but on responsible, synchronised integration. Below are the most promising application areas where AI is delivering real value across the MedTech lifecycle.

15.2.1 Diagnostics and Clinical Decision Support

AI's most visible impact has emerged in diagnostics, often as decision support, not decision replacement. Machine learning models trained on vast imaging and clinical datasets are now helping identify patterns the human eye might miss.

For example:

- In ophthalmology, AI systems flag diabetic retinopathy in seconds, increasing access in underserved areas.
- In cardiology, predictive models analyse ECGs to detect atrial fibrillation or cardiac arrest before symptoms arise.
- In oncology, radiomics algorithms uncover micro-patterns in scans that suggest malignancy even in early stages.

These systems don't replace clinicians. They elevate them, sharpening iagnostic accuracy and increasing speed, particularly in high-volume or resource-limited settings.

> *"The right AI doesn't diagnose for the doctor —*
> *it gives the doctor more to diagnose with."*

15.2.2 Predictive Maintenance and Field Health Monitoring

AI also plays a critical role behind the scenes, ensuring devices work reliably, safely, and cost-effectively throughout their lifecycles.

Key applications include:

- Predictive maintenance of factory equipment and calibration tools, preventing downtime.

- Embedded diagnostics in field-deployed devices that detect hardware degradation, software drift, or battery fatigue before failure occurs.
- Post-market complaint surveillance powered by AI, identifying signal patterns that would take months to surface through manual review.

This proactive insight reduces risk, protects patients, and prevents costly recalls or service events. In quality systems, AI becomes an early-warning engine, catching whispers before they become alarms.

15.2.3 Personalised Therapy and Adaptive Devices

The promise of personalised medicine, once elusive, is now tangible, thanks to AI-powered adaptive devices.

Examples include:

- Closed-loop insulin pumps that predict glucose spikes and adjust dosing automatically.
- Smart prosthetics that adapt to user gait and terrain, improving balance and comfort in real time.
- Neurostimulation implants that adjust signals dynamically based on real-time brainwave analysis.

These devices learn from their users, not just about them. AI turns passive instruments into active partners, continuously improving outcomes, comfort, and autonomy.

15.2.4 Operational Efficiency and Decision Automation

Beyond clinical use, AI is transforming how MedTech companies work internally, optimising R&D, compliance, and supply chain operations.

Emerging examples:

- AI tools that scan regulatory databases and flag changing requirements that impact in-progress submissions.
- Language models that draft Clinical Evaluation Reports or Risk Management Files, accelerating documentation cycles.
- Machine vision systems that review imaging data for completeness, reducing rework.
- Predictive supply algorithms that anticipate component shortages or demand spikes before they occur.

This is not just automation, but it is intelligent orchestration. AI liberates expert time from repetitive tasks, allowing focus on judgment, innovation, and quality oversight.

15.2.5 Human-Centric Design and Care Enablement

Perhaps AI's greatest contribution isn't what it automates, but what it enables for humans. By taking on data-heavy, repetitive tasks, AI frees clinicians to connect more meaningfully with patients, engineers to focus on design quality, and QA teams to act on insights, not just reports.

But this human enablement must be intentional. Systems must be transparent, interpretable, and aligned with real-world workflows. The most successful AI implementations in MedTech are the ones that feel intuitive, not intrusive, to their users.

AI, at its best, disappears into the background. It makes good decisions easier to see, risks easier to anticipate, and care more precise without overwhelming the people delivering it.

15.2.6 The Integration Imperative

AI's power doesn't lie in any single breakthrough. It lies in its integration: across systems, functions, and teams. Diagnostics improve when linked to post-market surveillance. Supply chain foresight sharpens when paired with complaint data. Adaptive devices get smarter when they learn from clinical outcomes.

This is the shift: from standalone AI pilots to synchronised intelligence ecosystems. The organisations that succeed won't be the ones with the flashiest algorithms. They will be the ones with the cleanest data, clearest oversight, and most connected teams.

The next section explores a complementary innovation, digital twins, and how they take the idea of learning systems even further, creating virtual environments to test, improve, and personalise devices before a single patient is touched.

15.3 Digital Twins: Prototyping Reality Virtually

> *"In the era of digital twins, quality is no longer something you test at the end — it is something you build from the very beginning."*

Digital twins represent one of the most transformative frontiers in MedTech. No longer just digital blueprints, they are dynamic, data-driven replicas of devices, organs, or even entire patients: learning, evolving, and simulating real-world performance in ways traditional methods never could.

5.3.1 What Are Digital Twins?

A digital twin is not simply a 3D model or simulation. It is a living, evolving digital mirror of a physical system, built to learn from real-world data and simulate how that system behaves, performs, and changes over time.

A robust digital twin integrates:

- High-fidelity models (physics-based, AI-augmented, or both)
- Real-time data from sensors, wearables, imaging, and EHRs
- Simulation engines that model biological and mechanical behaviour
- Validation loops to ensure predictions match reality

Unlike traditional simulations, digital twins operate continuously. They don't just represent, they reason.

15.3.2 Use Cases Across the Lifecycle

Digital twins are already transforming every phase of the MedTech lifecycle:

1. Design and Prototyping: Orthopaedic firms simulate years of wear on a virtual hip implant using thousands of gait variations, identifying weak points before the first prototype is built.
2. Testing and Verification: A cardiovascular device company runs virtual simulations of blood flow and turbulence in

different anatomical configurations, revealing potential complications long before animal or human trials.

3. Manufacturing and Post-market Surveillance: Field devices equipped with sensors feed real-time performance data into their twins. If performance drifts from the norm, predictive alerts are triggered, enabling intervention before failure.

4. Clinical Decision Support: Patient-specific digital twins are emerging in ICU settings, modelling respiratory or cardiovascular responses minute by minute, helping clinicians anticipate deterioration and adjust treatment dynamically.

5. Virtual Trials: Digital twins of diverse patient populations enable in silico testing of therapies, simulating outcomes across thousands of virtual patients to inform study design and accelerate approvals.

These use cases aren't speculative, they are already being applied by leading organisations to de-risk decisions, reduce time-to-market, and personalise care.

15.3.3 Validation and Regulatory Considerations

For digital twins to gain regulatory and clinical trust, they must meet stringent validation standards. It is not enough for a model to be plausible, it must be proven.

Key requirements include:

- Model validation: Does it reliably predict real-world performance?
- Data quality and lineage: Are inputs comprehensive, current, and representative?

- Version control: Can changes to the model be tracked over time?
- QMS integration: Are simulation outputs embedded in formal risk, design, and V&V processes?

Regulatory agencies like the FDA are actively supporting this evolution. During COVID-19, in silico models helped secure emergency use authorisations. New frameworks like Model-Informed Drug Development (MIDD) and Digital Evidence guidance are paving the way for broader adoption.

15.3.4 Building Digital Discipline

To realise the full potential of digital twins, organisations must move beyond experimentation to disciplined operational integration. That means:

- Treating virtual validation as rigorously as physical testing
- Training teams to understand simulation accuracy, bias, and applicability
- Embedding digital twin governance into QMS, design control, and risk management
- Establishing cross-functional ownership spanning R&D, QA, clinical, and regulatory

A digital twin is only as valuable as the trust it inspires, and trust comes from traceability, transparency, and repeatability.

15.3.5 The Strategic Impact

In MedTech, digital twins change more than development processes, they shift business models. When quality is built and tested virtually, teams move faster, learn earlier, and iterate safer. The result is not just speed, it is confidence.

The future of MedTech quality may be less about testing what already exists and more about simulating what might happen. And in that future, virtual twins will often be the first line of defence.

15.4 Challenges and Guardrails: Practising Responsible Innovation

"It is not enough to innovate. In MedTech, innovation must be explainable, equitable, and safe — by design."

Artificial Intelligence and digital twins bring unprecedented promise to MedTech, but also new dimensions of risk. These technologies don't just automate, they learn, evolve, and influence decisions in ways that traditional tools never could. That is why innovation without responsibility is not progress, it is peril.

Responsible innovation isn't a constraint. It is the enabler of trust, safety, and scalability.

15.4.1 Explainability and Transparency – Making AI Understandable

In healthcare, black-box models are not acceptable. Clinicians, patients, and regulators need to understand not just what a system concludes, but why.

Key principles include:

- Traceable logic: AI systems must articulate how inputs lead to outputs.

- Clinical interpretability: Outputs must align with medical reasoning, not just statistical accuracy.
- Human override: AI should inform, not dictate, with clinicians always in control.

Explainable AI (XAI) is now a regulatory and ethical baseline. A diagnosis without rationale isn't just unhelpful, it is unsafe.

"Explainability is not a bonus. In healthcare, it is a requirement."

15.4.2 Bias and Fairness – Designing for Every Patient

AI learns from data and data often carries the biases of its origin. If uncorrected, those biases can translate into clinical inequity.

Organisations must:

- Train models on diverse, representative datasets
- Conduct bias audits early and often
- Report outcomes stratified by age, gender, ethnicity, and comorbidities

Consider a digital health app that underperformed in rural populations because its training data was urban-heavy. Fairness isn't an afterthought, it is engineered from the ground up.

15.4.3 Safety Risk Management – Controlling a Moving Target

Unlike static devices, AI systems evolve. They update, adapt, and sometimes drift, meaning risk is no longer fixed at release, but must be monitored continuously.

Robust AI risk management includes:

- Model drift detection and alerting
- Real-time performance dashboards tied to PMS and CAPA
- Confidence thresholds, fallback mechanisms, and human-in-the-loop design

Traditional FMEAs are insufficient alone. Dynamic systems require dynamic controls, continuously updated and embedded into the QMS.

15.4.4 Regulatory Readiness – Meeting a Moving Standard

Regulators are adapting rapidly to AI and digital technologies, but they are asking hard questions.

Global developments include:

- FDA: Released the foundational Good Machine Learning Practice (GMLP) guiding principles, co-authored with Health Canada and UK MHRA, to inform safe and effective AI/ML-based SaMD development.
- IMDRF: Adopted and expanded the GMLP framework, shaping its principles into international harmonization tools for medical device regulation.
- EU: The proposed AI Act and ongoing MDR amendments add algorithmic transparency, risk management, and human oversight requirements.

Proactive strategies matter:

- Submit predetermined change control plans (PCCPs)
- Provide explainability and validation evidence during review
- Build regulatory literacy into product teams

Regulatory trust is not won with speed. It is won with structure, clarity, and readiness.

15.4.5 Guardrails as Accelerators, Not Obstacles

It is a myth that regulation slows innovation. In reality, good governance accelerates innovation by building the trust that enables scale.

The new standard for responsible AI in MedTech:

Attribute	Expectation
Transparent	Clinically explainable and regulator-ready
Fair	Trained and validated across diverse populations
Safe	Monitored, drift-controlled, and fail-safe
Governed	Embedded into QMS with traceability and lifecycle oversight

In MedTech, trust is not optional. It is your license to operate and your passport to leadership.

15.4.6 Culture of Responsibility – Beyond Checklists

The most critical guardrail isn't technical, it is cultural.

Organisations that lead in responsible innovation don't treat ethics as a compliance function. They make it a shared mindset. This includes:

- Empowering teams to speak up on fairness, safety, and transparency
- Making ethical risk part of design and review conversations

- Celebrating responsible innovation as a measure of quality, not friction

The leap into AI and digital systems isn't just technical. It is cultural. And culture is what sustains trust when things go wrong.

15.5 Embedding AI: Synchronising QMS and Culture

Artificial Intelligence and digital twins are not standalone features, rather they demand a rethinking of how MedTech organisations operate. Innovation must be synchronised not just across systems, but across teams, roles, and values. The future belongs to those who embed intelligence into the DNA of their quality systems and organisational culture.

15.5.1 Governance and Ownership – Who Owns the Algorithm?

Traditional roles and responsibilities in MedTech don't naturally accommodate learning systems. AI introduces fluidity, in performance, accountability, and change management.

To address this, forward-leaning organisations define clear roles:

- AI Product Owner: Oversees model lifecycle, retraining plans, and update governance.
- Data Steward: Ensures quality, provenance, and representativeness of training datasets.
- Digital Validation Lead: Integrates AI performance verification into the QMS.
- Clinical AI Liaison: Aligns model design with clinical realities and ethical expectations.

Without clear ownership, innovation risks becoming unmanaged and untrusted. Governance must be proactive, cross-functional, and continuous.

15.5.2 Measurable Validation – Moving Beyond One-Time Checks

AI performance isn't static, so validation can't be either. Instead, leading companies are embedding "living" validation systems that evolve with the model.

Best practices include:

- Real-time dashboards for performance indicators such as AUROC, sensitivity, and calibration.
- Drift detection protocols comparing current inputs to training data baselines.
- Retraining triggers tied to performance thresholds or post-market surveillance insights.
- Version control for model updates, with full traceability to risk management and clinical impact.

Validation is no longer an event before launch, it is a discipline that continues every day the model is in use.

15.5.3 Quality System Integration – AI as a Native Citizen of QMS

To synchronise innovation, your QMS must treat AI models like any other critical system component, subject to change control, risk review, and post-market vigilance.

Elements of a modern AI-ready QMS include:

- Inclusion of algorithm lifecycle in design controls (inputs, verification, validation, release).
- Integration with complaint handling, CAPA, and PMS processes for AI-driven systems.
- Specific SOPs for data versioning, model retraining, and explainability documentation.
- Audit readiness tools that track model decisions, updates, and rationale across time.

When AI is embedded into the QMS, it becomes part of the product's safety and performance backbone, not an afterthought or isolated experiment.

"In AI, safety is not a box to check once. It is a relationship to manage forever."

15.5.4 Upskilling the Organisation – A Workforce Ready for AI

You can't govern what you don't understand. The rise of intelligent systems requires an equally intelligent workforce with not just data scientists, but every function learning how to interact with, audit, and guide AI.

Successful companies are:

- Training QA/RA professionals in algorithmic auditing and bias detection.
- Educating clinicians on interpreting model outputs and uncertainty metrics.

- Equipping regulatory teams with AI-specific guidance and submission frameworks.
- Creating AI literacy pathways for all employees, from product managers to support engineers.

Some are even establishing internal academies, simulation labs, or "AI sprints" to co-create and learn together. When knowledge is synchronised, so is innovation.

15.5.5 Cultural Synchrony – Aligning Purpose and Practice

Ultimately, technology doesn't synchronise teams, culture does.

World-class organisations ensure that:

- Innovation is pursued with clarity of purpose – improving care, not chasing novelty.
- Ethical questions are asked early, not in crisis, but in design.
- AI performance is owned by cross-functional teams, not just engineers or vendors.

This cultural alignment creates resilience. When a model underperforms or a bias emerges, teams respond not with finger-pointing, but with shared accountability and agility.

AI becomes not just a feature, but a reflection of the company's values, discipline, and vision.

In the next section, we turn to one of AI's most transformative applications: virtual trials and simulated evidence. How do you test a product before it is even built? Let us explore how in silico innovation is reshaping the future of clinical validation.

15.6 Virtual Trials and Simulated Evidence: Accelerating Innovation

"We used to test on patients to protect them. Now, we can test before the patient ever enters the room."

Virtual trials or in silico simulations represent one of the most powerful leaps in clinical science. By using validated digital twins and predictive models, organisations can explore risks, optimise designs, and generate evidence: faster, safer, and more inclusively than traditional trials alone.

These simulations are not a replacement for real-world trials. They are a complement, a force multiplier for insight, safety, and speed.

15.6.1 Ethical and Strategic Advantages

Virtual trials offer a paradigm shift in both ethics and efficiency:

- They reduce patient exposure to unproven therapies by identifying failure modes early.
- They enable simulations across diverse demographic and physiologic profiles, overcoming recruitment limitations.
- They allow rapid iteration of dosing strategies, device parameters, and trial designs before costly human trials begin.

A neurostimulation company created digital brain twins based on diverse population data. These simulations revealed optimal stimulation profiles, accelerating first-in-human trials by over 30%, while avoiding multiple protocol amendments.

15.6.2 Use Cases and Applications

Virtual trials can be deployed in a wide range of MedTech applications:

- In Silico Prototyping: Test device performance under varied anatomical, usage, or disease states.
- Synthetic Control Arms: Reduce or eliminate placebo groups in trials for rare diseases or ethically sensitive conditions.
- Protocol Optimisation: Simulate different trial designs to find the best balance of safety, statistical power, and recruitment feasibility.
- Safety Edge Case Testing: Explore edge conditions, such as device misconfiguration, comorbidity interactions, or extreme physiologies, that are too rare or risky to test live.

These applications unlock safer, smarter, and more inclusive innovation.

15.6.3 Statistical Integrity and Validation

To gain trust, virtual trial data must meet rigorous scientific standards. This means treating digital evidence with the same scrutiny as traditional clinical data.

Key principles:

- Model transparency: Clearly define the underlying assumptions, boundaries, and limitations of the simulation.
- Validation against real-world data: Ensure digital twin predictions are benchmarked against existing clinical results or experimental studies.
- Population representation: Use diverse datasets to simulate outcomes across sex, age, ethnicity, comorbidities, and more.

- Regulatory-grade documentation: Maintain full traceability of inputs, model evolution, and output interpretation.

Done well, simulated evidence can complement physical trials or reduce their size, duration, and risk exposure.

15.6.4 Regulatory Readiness

Regulators are increasingly receptive to simulation-based evidence, provided it is delivered with rigour, clarity, and relevance.

Examples of global progress:

- FDA: Actively promotes model-informed development through its MIDD initiative and accepts virtual data in certain submissions.
- EMA: Embraces simulation in complex indications, such as paediatrics or rare diseases.
- IMDRF: Developing frameworks for digital evidence and model-based evaluation.

Best practices for regulatory success include:

- Submitting a Pre-Submission to align on modelling assumptions and protocols.
- Providing traceable justifications for every parameter and patient simulation.
- Including risk mitigations for model uncertainty or generalisability concerns.

Regulators don't expect perfection, they expect transparency, traceability, and readiness to adapt.

15.6.5 Internal Integration – Virtual Evidence as a System Capability

For virtual trials to scale, they must become part of the MedTech ecosystem, not an isolated experiment.

This requires:

- Cross-functional governance: Joint ownership across Clinical, Quality, Regulatory, and R&D functions.
- QMS inclusion: Virtual evidence must follow the same lifecycle controls as traditional data: versioning, change management, and validation.
- Knowledge reuse: Simulation results should feed into design history files, V&V strategies, and post-market surveillance frameworks.

Leading companies are developing "simulation libraries," reusable, validated digital cohorts that inform multiple projects over time.

15.6.6 The Opportunity Ahead

Virtual trials are not just a faster way to get to market. They are a smarter way to learn, to reduce harm, to increase inclusion, and to build evidence that would otherwise take years or millions to obtain.

When combined with a strong QMS and responsible innovation framework, virtual trials become more than a tool, they become a strategic advantage.

15.7 Embracing the Leap – Responsibly

The age of AI, digital twins, and virtual evidence is not just approaching, it has arrived. What was once theoretical is now powering operating rooms, guiding diagnostics, and shaping regulatory strategy. Yet with all its power, this leap forward demands discipline, humility, and purpose.

15.7.1 The Dual Imperative: Possibility and Accountability

These technologies grant us an extraordinary opportunity:

- To simulate thousands of patients before enrolling one.
- To identify failure before it happens, not after.
- To personalise therapy not just to a disease, but to a moment in a single life.

But alongside this promise comes new forms of risk. Bias, model drift, ethical ambiguity, and regulatory uncertainty are not footnotes, they are central design considerations. In MedTech, innovation isn't virtuous by default. It must be earned, every time, through transparency, validation, and stewardship.

15.7.2 From Vision to Action: Start Small, Scale with Intent

No organisation needs to transform overnight. But every organisation must start.

Begin with deliberate, high-impact steps:

- Pilot one explainable, bias-audited AI system in a clinical or operational context.

- Develop a validated digital twin of a known system to test V&V readiness.
- Run a virtual trial scenario under a defined QMS protocol to test governance.

Think of these as AI "learning loops," not one-time initiatives. Let outcomes, not ambition alone, determine your scaling path. And allow your quality system to serve not as a bottleneck, but as the engine of trust and acceleration.

15.7.3 Embedding Synchrony: The Culture That Powers Confidence

Throughout this book, we have explored how synchronisation across functions, systems, and mindsets unlocks sustainable excellence. In the context of AI and simulation, that principle becomes even more vital.

The leaders of tomorrow will be those who:

- Synchronise engineering with ethics and algorithms with assurance.
- Embed digital fluency into every corner of their workforce, not just the AI team.
- Treat responsibility not as a speed bump, but as the springboard for scale.

Innovation without integration creates risk. Integration without innovation stagnates. Great organisations achieve both, in rhythm.

15.7.4 This Is Not the End – It Is the New Beginning

This is not just a call to reflect. It is a call to leap, responsibly. Emerging technologies demand not just imagination, but intention. Ask yourself and your teams bold, practical questions:

- Where does AI currently intersect with your product or process, and is it embedded strategically or experimentally?
- Are your teams prepared to evaluate and manage AI-specific risks, such as algorithm drift, data bias, or transparency?
- How might digital twins or virtual evidence reshape your design, testing, or regulatory pathways?
- What cultural shifts are needed in your organisation to embrace AI with both speed and responsibility?

As we close this chapter, we don't conclude the conversation. We accelerate it.

> *"Innovation without integration is chaos.*
> *Integration without innovation is stagnation.*
> *True leadership demands both."*

The leap into AI, digital twins, and virtual trials is only the beginning. What follows is even more powerful: turning real-world data into continuous decision-making. Monitoring doesn't stop at market entry. Evidence doesn't end with approval. Learning doesn't pause between releases.

The next chapter takes us there, to the frontier of Real-World Evidence (RWE), post-market maturity, and the evolution of quality systems into living, learning engines. We will explore how data becomes decisions and how the best MedTech organisations are mastering the full product lifecycle, not just for compliance, but for clinical and commercial advantage.

Let us turn the page to Chapter 16 and discover how the learning never stops.

FROM DATA TO DECISIONS – RWE, VIRTUAL TRIALS, & POST-MARKET MATURITY

For decades, the finish line in MedTech seemed obvious: regulatory approval and product launch. But in today's data-rich, connected world, that finish line has shifted. Approval is no longer the end, it is the beginning of something far more valuable: continuous, real-world learning.

Every device placed in the field, every app downloaded, and every patient-clinician interaction generates real-world evidence (RWE). These signals (heart rate trends, usability feedback, device anomalies) offer far more than operational detail. They are the keys to safety, innovation, and trust.

This chapter explores how leading organisations are evolving from passive surveillance to post-market maturity. We will examine how RWE informs design improvements, how virtual trials are redefining clinical validation, and how synchronised teams are turning insight into impact. The goal is not just to monitor performance, but to improve continuously with intention, agility, and foresight.

In a world where the real test begins after launch, those who learn fastest will lead longest.

16.1 The Goldmine of Real-World Data

16.1.1 The Launch Is Just the Beginning

A cardiac monitoring device launched with strong trial data and wide acclaim. But months later, an anomaly emerged, an uptick in arrhythmia alerts among elderly users in a specific region. Investigation revealed a sensitivity calibration issue not captured in the original trial. Thanks to post-market vigilance, the issue was corrected before harm occurred.

This is the promise of real-world data: discovering what controlled trials cannot anticipate and acting before small issues scale into systemic risk.

Historically, product launch signified success. Today, it marks the start of a new phase, where clinical performance meets the complexity of everyday use. Devices don't just perform, they emit signals, generate feedback, and reveal hidden insights. Learning from this data is not optional, it is essential to safety, differentiation, and sustained innovation.

16.1.2 A New Era of Evidence

Clinical trials remain foundational, but their limits are increasingly clear. They focus on ideal conditions, not typical patients. They often exclude the elderly, the rural, the chronically ill. As a result, they offer control but not completeness.

Real-world evidence (RWE) bridges this gap. Captured from device telemetry, EHRs, apps, claims data, and registries, RWE shows how products function outside the clinical bubble, in the diversity of real life. It reveals long-term performance, rare interactions, usability challenges, and unmet needs.

Used well, RWE can:

- Refine inclusion criteria and training strategies
- Detect early safety signals
- Support iterative product updates
- Inform supplemental regulatory filings

What once lived at the edge of regulatory acceptability is now central to decision-making. RWE has shifted from anecdote to advantage.

16.1.3 From Signals to Strategy – Listening as a Capability

Synchronised organisations treat real-world data not as a compliance exercise, but as strategic infrastructure. They build systems that listen continuously, learn quickly, and act decisively.

Consider a neuromodulation firm that invited patients to log daily quality-of-life data through a companion app. After a year, patterns emerged, revealing that a specific therapy mode was more effective in post-menopausal women. This insight sparked a firmware update, a messaging adjustment, and a new clinical hypothesis.

This wasn't just reactive monitoring. It was a philosophy with patients as co-creators, the field as a living lab, and each data point as a clue to improving outcomes.

To lead in this environment requires more than dashboards. It requires humility, curiosity, and a system-wide commitment to learn from the real world, not just document it.

"In a data-rich world, ignorance is no longer a
lack of information. It is a failure to listen."

16.1.4 From Collection to Action

Data is only valuable when it leads to action. A mature post-market system doesn't simply collect data, it transforms it into meaningful improvement.

Ask:

- Are your data systems designed to detect weak signals early?
- Do you have a cross-functional team ready to act when trends appear?
- How fast can you go from detection to intervention?

In a data-rich world, the gap between what is known and what is done defines maturity.

Let us now explore how this shift toward real-world learning transforms clinical validation, trial design, and long-term product performance.

16.2 Real-World Evidence: Beyond the Clinical Trial

16.2.1 The World as the New Laboratory

Clinical trials are foundational, but they offer a curated, narrow view, highly controlled settings, selective populations, and short timeframes. They tell us what is possible, but not always what is probable in the real world.

Real-World Evidence fills that gap. It comes from electronic health records, device telemetry, insurance claims, patient-reported outcomes, mobile apps, and connected wearables. It shows how products perform across diverse geographies, conditions, and care settings.

RWE shifts the central question from "Can this work under perfect conditions?" to "Does this work for the people who actually use it?"

Used well, RWE brings clinical relevance, population insight, and environmental realism, transforming our understanding of performance, equity, and unmet needs.

16.2.2 The Regulators Are Listening

What was once viewed as anecdotal is now gaining regulatory traction.

In the U.S., the FDA's guidance on Use of Real-World Evidence to Support Regulatory Decision-Making for Medical Devices has laid the groundwork for RWE's integration into submissions, label expansions, and post-market follow-up. In Europe, the EMA's DARWIN EU initiative is establishing a pan-European data network to inform regulatory and health policy decisions.

Beyond these regions, the International Medical Device Regulators Forum (IMDRF) is aligning global perspectives, focusing on reliability, traceability, patient privacy, and governance standards.

This is more than procedural evolution, it is a validation of a broader, more inclusive evidence model. Regulators increasingly want to know not just what worked in a trial, but what works in the wild.

"We are not just interested in how a device performs
in the ideal world — we want to know how it performs
in the real world, where lives depend on it every day."

16.2.3 Building a Living RWE System

RWE is only as powerful as the system that supports it. Mature organisations design integrated RWE systems that go far beyond passive data collection.

Core components include:

- Data Infrastructure – Secure pipelines connecting EHRs, device telemetry, patient registries, mobile platforms, and claims data.
- Analytical Intelligence – Statistical tools, cohort models, AI/ ML engines, and longitudinal dashboards to find meaning in the noise.
- Governance - Clear rules for consent, data integrity, and compliance, aligned with global regulatory frameworks

This isn't just about data pipelines, rather it is about organisational metabolism. Fast learners outpace slow reactors.

Data must be curated, contextualised, and transformed into insight that informs not just Quality or Regulatory, but R&D, Clinical, and Commercial.

"Data without action is noise. Insight
without courage is wasted potential."

16.2.4 From Insight to Iteration – Evolving with the Evidence

When RWE is connected to action, it becomes an engine of evolution.

Consider a vascular implant manufacturer who discovered, through registry analysis, a higher rate of restenosis in patients over 80. Rather than dismissing it as a statistical anomaly, the company dug deeper. Physiological shifts in vascular elasticity were affecting product performance. Within months, a geriatric-optimised stent was in design. Because the real-world signal was actionable, the update was efficient and regulatory dialogue was accelerated.

This is what it means to move beyond surveillance:

- Use RWE to explore sub-population insights
- Refine design based on long-term, large-scale use
- Adjust training and labelling based on emerging risk profiles
- Proactively guide future regulatory and market strategies

Real-world evidence doesn't replace trials, it completes them. It validates in diversity what trials confirm in control.

16.2.5 Strategic Advantage, Ethical Imperative

The era of RWE is not just an opportunity, rather it is a responsibility. If the tools exist to understand performance more clearly, across more people, in more places, then failure to use them is not just a technical oversight. It is an ethical one.

RWE helps us reach underrepresented populations. It reduces blind spots. It empowers faster, more adaptive learning. And it extends product lifecycle through intelligent iteration.

When embraced fully, RWE turns compliance into confidence, and monitoring into momentum.

As we move into the next section, we explore how the RWE mindset reshapes trials themselves, into hybrid, decentralised, and increasingly virtual models that match how real people live and engage with care.

16.3 Virtual Trials and Hybrid Studies: Redefining Validation

16.3.1 From Controlled to Connected

Traditional clinical trials were designed for control: fixed locations, standardised protocols, and tightly managed inclusion criteria. This rigor brought precision, but also limitations. Recruitment was slow, participation narrow, and real-life relevance often missing.

Today, digital tools, remote monitoring, and patient-centred design are dissolving those boundaries. Trials are no longer limited to brick-and-mortar sites. They are decentralised, hybrid, and increasingly virtual.

This transformation isn't just logistical, it is philosophical. It shifts trials from controlling the patient experience to meeting patients where they are.

A MedTech company piloting a virtual heart failure trial enabled patients to enrol from home, upload vitals via Bluetooth, and connect with clinicians via video. The result: 50% faster recruitment, a broader demographic mix, and better retention.

> *"Virtual trials don't dilute scientific rigor, they extend its reach."*

16.3.2 Expanding Access, Deepening Equity

Decentralised and hybrid trials open the door to populations long excluded from traditional research: rural patients, the elderly, those with limited mobility, caregivers, and communities historically underrepresented in health data.

Virtual participation removes logistical friction – travel, time off work, site availability – and replaces it with convenience, inclusivity, and dignity.

By bringing trials to patients, rather than requiring patients to fit into trials, we unlock a deeper truth: equity isn't just about outcomes, it starts with access.

16.3.3 Adaptive Design and Virtual Controls

The digital transformation of trials enables a shift not only in where data is collected, but how trials are structured.

With adaptive design, real-time data flows inform mid-trial decisions: adjusting dosage, reallocating patient arms, or halting ineffective cohorts. Virtual dashboards guide these decisions dynamically.

Simultaneously, virtual control arms, built from historical or synthetic datasets, reduce the need for traditional placebo groups, especially in rare diseases or high-risk interventions. These models improve enrolment speed, reduce patient burden, and enhance statistical robustness.

Adaptation and virtualisation make trials smarter, faster, and more ethical, without compromising scientific integrity.

16.3.4 In Silico Trials - Digital Twins, Real Impact

Digital twins, computational models of anatomy, physiology, or disease progression, are now part of modern clinical planning. These in silico environments simulate patient responses, mechanical interactions, and biological variability before a single device is implanted.

In one example, a structural heart device was evaluated on thousands of virtual patients with varying anatomies. This revealed optimal sizing ranges, led to algorithmic refinements, and reduced rework in the physical trial by 40%.

In silico trials don't replace human studies, they augment them. They reduce risk, accelerate learning, and improve design confidence ahead of regulatory review.

16.3.5 Case Study - Crisis as Catalyst

At the height of the COVID-19 pandemic, one wearable MedTech firm faced a dilemma: how to validate a respiratory monitor when clinics were closed and patient visits halted.

Their response:

- Devices shipped directly to patients
- Onboarding conducted via video and mobile app
- Continuous data uploaded via secure cloud telemetry

Not only did the trial meet its endpoints, but it also finished ahead of schedule, with broader patient inclusion and higher engagement. Regulators praised the design, and the company adopted the hybrid model as standard for future trials.

What began as a workaround, can became a breakthrough.

16.3.6 The New Validation Standard

Virtual and hybrid trials are no longer experimental, they are essential.

They don't simply replicate physical trials online, they redefine what validation looks like in a connected, real-time world. They integrate wearables, AI-based endpoints, voice-based symptom capture, and app-driven engagement.

Smart organisations aren't asking whether to adopt these trials. They are asking how to design them better to be more inclusive, more adaptive, more predictive.

This evolution isn't the end of clinical validation, but it is its reinvention.

16.4 From Post-Market Surveillance to Post-Market Maturity

16.4.1 From Passive Surveillance to Active Maturity

For decades, post-market surveillance (PMS) was a compliance checkbox: reactive, episodic, and event-driven. Its purpose was to identify and contain harm after it happened. But today, that paradigm is being redefined.

In a world of connected devices, real-time telemetry, and digital health ecosystems, leading organisations are transforming PMS into post-market maturity (PMM) – a continuous cycle of monitoring, learning, and improvement.

PMM reflects a fundamental mindset shift: the field is not merely a risk zone, but a second innovation lab. Every device in use becomes a sensor. Every signal, a potential source of design evolution.

High-maturity organisations no longer ask, "Are we compliant?" They ask, "What are we learning?"

16.4.2 Closing the Feedback Loop – From Insight to Action

True post-market maturity is about turning data into decisions.

Synchronised organisations build closed-loop systems that absorb signals from across the ecosystem – complaints, usage trends, telemetry, customer feedback – and act with agility.

Examples include:

- If performance declines in high-altitude environments, a firmware update adjusts parameters for local conditions.
- If UI friction is reported by field reps, usability improvements are designed and deployed proactively.
- If registry data reveals off-label use in a specific region, training is adapted to reinforce clinical boundaries.

The feedback loop is only closed when insight drives intervention and when those interventions are shared transparently across teams.

"The loop is not closed until feedback becomes design.
The cycle is not complete until vigilance becomes value."

16.4.3 Maturity is Measurable

Post-market maturity is not abstract, it is measurable.

World-class organisations track not just adverse events but the speed, precision, and effectiveness of their response. This has led to the definition of the Post-Market Maturity Index (PMMI), a structured approach to assessing organisational readiness and responsiveness.

Typical PMMI dimensions include:

- Signal detection latency
- Time-to-mitigation
- Volume of proactive vs. reactive updates
- Stakeholder engagement (clinician, patient, regulator)
- Integration of post-market data into design and roadmap

A maturity ladder helps visualise the progression:

Level	Descriptor	Characteristics
1	Reactive	Reports event, closes CAPAs slowly
2	Proactive	Tracks trend, surfaces weak signals
3	Adaptive	Updates designs and guidance based on feedback
4	Predictive	Anticipates issues through simulation and AI

This maturity isn't about passing audits, it is about accelerating trust, reducing surprises, and enabling better outcomes.

"A product's maturity is not defined by its age—but by how wisely it grows."

16.4.4 Earning Trust Through Engagement

Mature organisations don't treat PMS as a background activity, rather they bring it to the forefront of stakeholder engagement.

They notify clinicians about performance improvements. They involve patient advocates in usability reviews. They invite regulators into the learning loop early and transparently.

For example:

- When an Instructions for Use (IFU) update is released, it is accompanied by training modules and a thank-you note to those who flagged the gap.
- When telemetry data prompts a performance enhancement, field reps are informed and armed with talking points for customer reassurance.
- When usability feedback shapes an update, contributors are acknowledged internally and externally.

Trust is no longer a product of compliance. It is the result of visibility, honesty, and shared learning.

16.4.5 From Devices to Ecosystems

Post-market maturity extends beyond individual products, it spans portfolios, platforms, and partnerships.

Smart organisations are building integrated "listening systems" that cut across devices and regions, allowing them to:

- Compare performance across product generations

- Detect cross-platform user friction
- Align updates across global markets

For example, a diagnostics firm noticed a higher complaint rate for two different test kits across different geographies. Rather than treat them in isolation, they found a shared usability flaw in the mobile reader interface and addressed both with a single harmonised update.

This level of coordination creates resilience at scale. It turns feedback into foresight and isolated issues into enterprise intelligence.

16.4.6 Post-Market Maturity as a Strategic Advantage

The best organisations don't view post-launch life as a maintenance phase. They see it as the core of their learning engine.

Post-market maturity:

- Extends product lifecycle through continual refinement
- Builds brand trust through transparency and responsiveness
- Reduces recall risk and strengthens regulatory relationships
- Enables faster iteration and smarter innovation pipelines

It transforms vigilance into value creation and repositions the quality function from a safety monitor to a strategic partner.

As we move into the next section on organisational culture and decision-making, one truth becomes clear:

> *"The future belongs to those who don't just detect risk but learn from it, act on it, and share that learning across the enterprise."*

16.5 Building a Culture for Data-Driven Decisions

16.5.1 Cultivating the Learning Organisation

Building a culture of evidence-driven decisions is not about having more data, it is about creating an environment where data becomes dialogue and insight drives action. The highest-performing MedTech companies don't wait for quarterly reviews or retrospective analysis. They operate in a state of continuous learning.

This culture starts with leadership. Leaders must actively champion transparency, curiosity, and responsiveness. In a learning organisation:

- Issues are surfaced early, not hidden
- Data is shared cross-functionally, not siloed
- Responses are collaborative, not reactive

A company that once viewed post-market signals as threats now holds regular multi-disciplinary "insight huddles" where quality, clinical, commercial, and regulatory teams assess signals in real-time and co-design responses. The result? Faster iteration, stronger cross-team cohesion, and better patient outcomes.

> *"Data doesn't build trust. How we respond to data does."*

16.5.2 Infrastructure That Connects and Amplifies

Even the most data-literate culture cannot thrive without the right infrastructure. Leading organisations invest in unified, intelligent systems that bring together data from diverse sources: complaints, telemetry, clinical feedback, CRM tools, and field reports.

Key enablers include:

- Federated data platforms for secure, cross-functional access
- Unified dashboards that combine technical, clinical, and business insights
- Real-time alerting systems that flag deviations before they become failures

A diagnostics firm created a digital "control tower" that linked post-market data with development roadmaps and risk management plans. With one shared interface, teams could track trends, simulate impact, and assign ownership, all within hours of a new signal emerging.

> *"It is not the quantity of data that makes the difference, it is the coherence."*

16.5.3 People Who Translate Signals into Strategy

Technology can deliver information, but it is people who extract meaning.

To thrive in a data-driven world, MedTech organisations must develop new roles and evolve existing ones. These include:

- RWE Analysts who interpret longitudinal data from registries, claims, and wearable streams
- Clinical Translators who bridge usage insights with design decisions
- Regulatory Informatics Leads who ensure new evidence streams align with compliance requirements
- Digital Stewards who safeguard data integrity and reliability

But the transformation doesn't stop at these roles. Every function (engineering, quality, marketing, supply chain) must become more data-literate. They must be able to ask better questions, spot early signals, and translate findings into action.

"The goal is not data fluency in isolation, rather it is organisational fluency in insight."

16.5.4 Systems That Drive Response, Not Just Awareness

Insight is only valuable when paired with structure and accountability. That is why leading companies establish formal response mechanisms to act on the signals they detect.

Common examples include:

- Post-Market Intelligence Councils that meet monthly to review emerging data and prioritise responses
- Signal-to-Action Playbooks that map decision pathways, escalation triggers, and communication workflows
- Product Evolution Boards that link field insights to roadmaps and next-gen development

For instance, a wearables company detected an unusual spike in false alarms among adolescent users. Instead of dismissing the anomaly, their cross-functional response group investigated, discovered a sensor-skin interaction variance, and deployed a patch, backed by updated guidance and field training, within two weeks.

This wasn't just compliance. It was customer-centricity in motion.

16.5.5 Measuring What Matters – Trust, Speed, and Learning

The ultimate measure of a data-driven culture is not how much data you collect, but how quickly and wisely you act on it.

High-performing MedTech teams track:

- Signal detection latency
- Time-to-intervention
- Percentage of proactive versus reactive updates
- Stakeholder engagement in learning and response loops
- Impact of insights on future design or process improvements

The goal is not just accuracy, it is agility. Not just precision, but participation.

Because when an organisation listens actively, responds wisely, and learns publicly, it does more than manage risk, it earns trust. And in today's interconnected, transparent, regulated world, trust is the highest currency.

As we move into the final section of this chapter, we will explore how all these elements – systems, culture, and evidence – converge to build the ultimate MedTech capability: becoming a true learning organisation.

> *"Insight doesn't come from tools. It comes from teams that can ask better questions."*

16.6 The Learning Organisation

16.6.1 Beyond Launch – The New Lifecycle Mindset

For too long, launch was seen as the finish line, where development ended and maintenance began. But in the synchronised MedTech era, launch is just the ignition point. The real value begins once the device enters the real world.

Every real-world interaction is an opportunity to learn about patients, systems, and safety. The most mature organisations don't just collect data. They grow from it. They don't wait for complaints. They design updates based on anticipation, not escalation.

This mindset shift, from post-market surveillance to post-market evolution, is what defines a learning organisation. It is not a reactive function. It is a cultural commitment.

16.6.2 Evidence Over Assumption – The New Standard

As W. Edwards Deming famously said, "In God we trust. All others must bring data." But in the modern MedTech landscape, data isn't just a compliance tool, it is the foundation of innovation.

Clinical trials remain vital, but they are no longer sufficient. RWE adds depth, diversity, and durability. It fills the blind spots of pre-market studies, bringing equity, usability, and context into decision-making.

Learning organisations don't rely on hunches or hierarchy. They interrogate reality with humility. They use evidence to validate assumptions, refine designs, and improve outcomes, continuously.

16.6.3 Systematised Listening – Culture and Capability

Real-time telemetry, patient-reported outcomes, app analytics, and field feedback – these are not just data streams. They are voices.

The most advanced teams don't simply measure, they interpret. They embed listening into their governance structures, team rituals, and digital ecosystems. They invest in platforms that integrate data and in people who can turn signals into strategy.

In a learning organisation:

- Quality and clinical teams operate with shared dashboards
- Commercial teams use RWE to guide market strategy
- Regulatory teams proactively align signals with submission updates
- Engineers refine the roadmap based on what the field teaches

And most importantly, leadership treats learning as a measure of strength, not a symptom of failure.

16.6.4 A New Identity – From Manufacturer to Learning Partner

Post-market maturity doesn't just change what we do, it changes who we are.

A learning organisation is not just a manufacturer of devices. It is a partner in care. A system of listening. A living bridge between product and patient. It earns trust not by saying "we are compliant," but by proving, again and again, that it is thoughtful, responsive, and evolving.

This is how modern MedTech wins, not through speed alone, but through synchronisation. Not just through design excellence, but through insight agility. It is not about more data. It is about wiser action.

16.6.5 Call to Action - Operationalising the Mindset

The transformation from surveillance to intelligence doesn't happen overnight, but it must start intentionally.

This week, ask yourself:

- What is the most underused signal source in our post-market process?
- What is one insight we have observed but not yet acted upon?
- Where can we close the loop, faster, smarter, and more transparently?
- Are you treating real-world data as a goldmine or as a regulatory checkbox?
- How integrated is RWE into your decision-making across lifecycle stages—not just post-market?

Pick one signal. Start one conversation. Define one action.

The learning loop begins not with data, but with willingness.

As we conclude this chapter on real-world learning, one final shift comes into view: the move from insight to articulation. Because the best organisations don't just learn, they document their learning in a way the world can understand.

In the next chapter, we will explore how to translate all this real-world intelligence: clinical data, quality decisions, risk rationales, into global-ready documentation that doesn't just satisfy regulators but showcases maturity.

From dossier dread to digital clarity, from fragmented templates to harmonised narratives, we will show how MedTech leaders are reinventing the submission process and making documentation not suck.

Let us move from what we have learned to how we express it: with precision, clarity, and confidence.

CHAPTER 17

THE SUBMISSION SHIFT – TURNING COMPLIANCE INTO COMPETITIVE ADVANTAGE

E very MedTech journey reaches a defining moment: the point when invention must become evidence, and evidence must become approval. No matter how brilliant the device, it will not reach a patient until it passes through the regulatory gateway. And too often, that gateway is approached with dread.

Many companies still approach submission as the last step in development, missing the chance to use it as a powerful expression of their product's journey and integrity. The result is predictable: late nights, missed details, frustrated teams, and delayed access for patients who cannot afford to wait.

But submission does not have to be punishment. It can be power. When approached with foresight and discipline, it becomes more than a regulatory hurdle, it becomes a strategic advantage. A strong submission does not just satisfy regulators, rather it demonstrates maturity, earns trust, and accelerates global access.

This chapter shows how submission can shift from being a roadblock to becoming a pathway forward. We will look at the systems, behaviours, and cultural shifts that transform documentation into a clear, persuasive narrative of product excellence. We will examine the role of synchronisation across functions, the discipline of building for global readiness, and the tools that enable speed without compromise.

In MedTech, submission is not the end of innovation. It is its most visible expression, the point where what you built becomes what the world can believe in.

17.1 Reframing the Mindset: From Burden to Advantage

For decades, regulatory submission has been seen as the necessary evil of MedTech. Teams pour years of effort into design, clinical validation, and risk management, only to reach the final stretch and feel crushed by the weight of documentation. Submissions are often viewed as an administrative burden, a stressful scramble of compiling files, formatting reports, and filling gaps under pressure. This perception is not only unfair, but also deeply counterproductive. It diminishes morale, delays access to patients, and reduces submission to a bureaucratic exercise rather than what it truly is: the ultimate expression of product excellence.

The submission shift begins with a change in mindset. Submission should not be treated as the end of innovation, but as its stage. It is the moment to demonstrate the clarity of thinking behind a device, the robustness of the data, and the integrity of the organisation that stands behind it. When reframed in this way, submission transforms from a bottleneck into a strategic advantage.

> *"A scattered development journey creates a scattered dossier. A synchronised journey creates a persuasive one."*

17.1.1 From File Dump to Strategic Narrative

A submission is not a data dump. It is a biography of a product. Done poorly, it becomes a warehouse of disconnected documents that confuse reviewers and raise doubts. Done well, it reads as a coherent narrative: why the product was created, how risks were identified and mitigated, what evidence justifies its claims, and how patients will ultimately benefit.

This narrative is engineered, not improvised. It requires alignment across functions, from design history to risk files, from clinical evaluations to labelling. When these threads are woven into a clear story, the reviewer is guided rather than burdened. Instead of wading through noise, they see logic, coherence, and care. In this way, the submission becomes not only a regulatory artefact but a demonstration of organisational maturity.

17.1.2 Synchronisation as the Foundation

A fragmented development process produces fragmented submissions. Misaligned claims, inconsistent terminology, and duplicated evidence all signal a lack of coherence. Regulators see these gaps not simply as clerical errors, but as cracks in the system that produced the device.

Synchronisation is the antidote. When quality, regulatory, clinical, and engineering teams work together from the earliest stages, submissions emerge as a natural output of the process, not a desperate afterthought. Intended use, performance claims, and risk controls are aligned from design inputs onward. Documentation flows with consistency because the journey itself was consistent.

"Submission is not the job of regulatory affairs alone, but it is the culmination of cross-functional alignment. The tighter that alignment, the more persuasive the submission."

17.1.3 The Strategic Advantage

Why does this matter? Because a submission is far more than an entry ticket to market. It is a signal. To regulators, it signals whether the company is disciplined and trustworthy. To investors, it signals whether the organisation can scale. To internal teams, it signals whether their hard work is aligned into one coherent truth.

A strong submission does four things at once: it accelerates approvals, it builds credibility, it fosters internal clarity, and it demonstrates maturity. These outcomes are not "nice to haves." In a sector where speed, trust, and reliability are existential, they are competitive differentiators. A company that consistently delivers clean, credible submissions gains more than regulatory clearance, it earns respect, and respect compounds into long-term advantage.

17.1.4 A Mindset Shift Worth Making

Changing how organisations view submission is both a cultural and strategic shift. It requires moving from the language of "survival" to the language of "showcase." Instead of dreading the dossier, teams can take pride in it. Instead of fearing deficiencies, they can anticipate confidence.

This is the Submission Shift. It is not about adding more effort but about aligning effort differently. When submission is seen as the bridge between innovation and impact, it is no longer a painful epilogue. It is the peak moment when everything comes together, not just to gain approval, but to earn trust.

17.2 Building for Global Readiness: The Core Dossier Approach

For companies working in today's interconnected MedTech landscape, submission readiness can no longer be a local exercise. The traditional pattern – file with the FDA first, then modify for CE marking, and later rework for Canada, Australia, or Japan – is not only inefficient, but also unsustainable. It breeds inconsistencies, creates avoidable delays, and drains already stretched teams.

The way forward is not to work harder, but to work smarter. That means designing for global readiness from day one, using an approach that is structured, scalable, and capable of adapting to regional nuance without reinventing the wheel each time.

> *"If your product is built to cross borders, your documentation must be built to travel too."*

17.2.1 Core Dossier Thinking – One Story, Many Languages

The foundation of global readiness is the Core Dossier: a harmonised, modular documentation framework that acts as a single source of truth for all submissions. Instead of rewriting narratives for every jurisdiction, organisations must build a validated backbone (device description, intended use, design rationale, risk controls, clinical evidence) that can be flexibly adapted to local requirements.

This is not one overstuffed master file. It is a system of interoperable modules, written once with global rigour, then tailored for specific regions. When done well, a Core Dossier reduces duplication, accelerates subsequent filings, and strengthens both internal coherence and reviewer confidence.

17.2.2 Architecting for Reuse – Modular, Traceable, and Scalable

A Core Dossier begins with architecture. Leading organisations build documentation around globally recognised frameworks such as the IMDRF Table of Contents or Summary Technical Documentation. These structures ensure modularity, traceability, and scalability.

Key design elements include:

- Self-contained modules that can be updated independently without destabilising the whole.
- Version control and metadata tagging by geography, product version, and submission type.
- Reusable evidence sets (clinical, biocompatibility, usability) generated at the highest global standard and mapped to multiple jurisdictions.

With this approach, changes cascade automatically, consistency is maintained, and rework is dramatically reduced.

17.2.3 Global by Design – Anticipating Variability, Not Reacting to It

True global readiness means anticipating regional differences at the outset, not scrambling after the fact. Clinical protocols can be designed to satisfy both FDA endpoints and EU equivalence justifications. Risk

THE SYNCHRONISATION SHIFT

files can be aligned to ISO 14971 while mapping to FDA-specific expectations. Labelling and UDI systems can be structured to meet global nomenclature requirements while accommodating local detail.

This is not about making every market identical. It is about embedding flexibility so that adaptation is swift, deliberate, and reliable. Global-ready submissions signal respect for regulatory nuance while maintaining a coherent product story across borders.

17.2.4 The Operational Payoff – From Reactive to Agile

The advantages of a Core Dossier approach go beyond documentation efficiency. They reshape how organisations operate.

- Duplication across functions and geographies is reduced.
- Regulatory changes can be absorbed and implemented faster.
- Audit readiness is strengthened through consistency and traceability.
- Product, clinical, and regulatory teams align around a shared narrative.

Most importantly, the cultural tone changes. When teams write for a global audience, they hold themselves to a higher standard. The dossier becomes more than paperwork, it becomes an act of alignment, precision, and confidence.

17.2.5 Case in Point – The Strategic Dossier Backbone

A mid-sized cardiovascular company adopted a modular submission backbone built on the IMDRF Table of Contents. By embedding traceability from design history through post-market surveillance, and by training cross-functional teams to co-author within a single platform, they transformed their submission capability.

The outcome was striking. Their next-generation ECG device achieved CE mark and FDA clearance within six months of each other. Internal review cycles were cut in half. Regulatory feedback focused on scientific nuance, not documentation gaps. The Core Dossier became more than a compliance tool, it became a growth enabler.

17.2.6 Final Thought – Designed to Scale

A Core Dossier is not a formatting trick. It is a philosophy of readiness. It says: we expect to expand, we expect to scale, and we have built the systems to do so without compromise.

Global readiness is not about crossing more borders. It is about crossing them better, faster, and with greater trust. The Core Dossier is the architecture of that trust, the structural backbone of a company prepared to lead across markets.

17.3 Writing with Purpose: Clarity, Credibility, Coherence

A regulatory submission is far more than a repository of data. It is the product's first formal conversation with the world. And like any conversation that matters, how you say it matters as much as what you say.

Poorly written submissions undermine confidence. They confuse, contradict, and slow review cycles. Strong submissions, by contrast, elevate understanding. They speak clearly. They earn trust. They move decisions forward.

Writing with purpose is not about elaboration, rather it is about precision. It is about making your evidence legible, your reasoning transparent, and your story coherent.

"A submission is not just a compliance artifact. It is your product's first conversation with the world."

17.3.1 From Data Dump to Scientific Storytelling

Data alone does not persuade. Reviewers do not want a mountain of raw information, they want a structured narrative that explains why decisions were made, how evidence was generated, and what that evidence proves.

A strong submission follows a logical arc:

- What risk was identified?
- What actions were taken?
- What evidence was generated?
- What does that evidence support?

This sequence is not creative writing, it is scientific storytelling. It guides reviewers through your thinking, makes conclusions credible, and accelerates their ability to validate your claims.

17.3.2 Structuring for the Reviewer's Journey

Every submission is read by a human being, often under pressure, often juggling multiple devices at once. Respecting that reality means writing with empathy for their experience.

Practical principles include:

- Lead with claims, not caveats.
- Use clear signposts and summaries for navigation.
- Repeat only to reinforce, not to clutter.
- Keep terminology consistent across every document.

The goal is not to simplify the science but to make intelligence accessible. Submissions that guide reviewers clearly signal maturity and competence.

17.3.3 Avoiding Clarity Killers

Even capable teams fall into traps that damage clarity. Among the most common:

- Inconsistent terminology: "sensor" in one file, "probe" in another, "electrode" in a third.
- Circular justifications: "Risk is acceptable because it is mitigated. It is mitigated because it is acceptable."
- Unlabelled data: charts without legends, tables without units, appendices without references.
- Contradictions: claims in the IFU that don't align with the CER.

None of these errors require more budget to fix, but they do require more discipline. Addressing them not only smooths review but also strengthens organisational credibility.

17.3.4 Building a Culture of Peer Review

Great submissions are never the work of a lone author. They are collective achievements, shaped by the discipline and perspective of a team. The best organisations understand this and treat peer review not as an optional step, but as an essential part of their culture.

They bring in colleagues who are unfamiliar with the product to perform fresh "cold reads," testing whether the submission communicates clearly to someone encountering it for the first time. They stage review simulations where teams step into the shoes of regulators, interrogating claims and assumptions with the same scrutiny an external assessor would apply. And they rely on trained writers and editors to sharpen the language, rooting out ambiguity, inconsistency, and unnecessary complexity.

These practices elevate submissions from hurried collections of documents into polished, persuasive narratives. They don't just reduce deficiency letters or shorten review cycles, but they strengthen confidence across the organisation. Teams begin to take pride not only in the device they built, but in the story of how it was built and why it can be trusted. That pride, when carried into a submission, is as powerful as the data itself.

17.3.5 Communicating at Two Levels – Technical and Lay

Modern regulators expect submissions that speak to both experts and non-experts. With EU MDR's Summary of Safety and Clinical Performance, FDA patient summaries, and Health Canada readability requirements, dual-level clarity is no longer optional.

Smart teams go further. They draft patient-level summaries first. Doing so forces them to distil product value, risk, and performance into plain language. This discipline sharpens both the lay narrative and the technical dossier that follows.

The ability to explain your device to a reviewer and to a patient is not just writing skill, it is organisational maturity.

17.3.6 Final Thought – Writing is Leadership

A submission does more than present data; it reveals how an organisation thinks. When writing is clear, it signals discipline. When reasoning is transparent, it signals integrity. When terminology is consistent, it signals control.

Strong submissions are not just accurate, they are accessible. They are purposeful. They say, with quiet authority:

"We know what we are doing, and
we want you to see it."

17.4 Operational Excellence: Systems, Tools, Traceability

Submission excellence does not begin at the keyboard, it begins with systems. Words and structure matter, but they rest on a foundation of infrastructure: how information is created, managed, reused, and validated. In high-performing organisations, operational excellence in submission is not about working longer hours. It is about working within systems that remove friction, reduce duplication, and embed rigour into every step.

Documentation is not simply a compliance artefact. It is a mirror of organisational maturity. The way content is managed tells regulators as much about a company's discipline as the data itself. In this sense, systems are not back-office tools but enablers of trust.

*"Your documents shouldn't contradict themselves —
and they don't have to, if your tools talk to each other."*

17.4.1 From Chaos to Coherence

Too many organisations still run submissions in a state of controlled chaos. Documents live in scattered folders, hidden on personal drives or locked in outdated systems. Version control is done manually, often with spreadsheets or endless email chains, leaving teams unsure which version is truly final. Hours are wasted reconciling inconsistencies, while deadlines loom and confidence erode. This is not a failure of talent, as the people are often brilliant and dedicated, it is a failure of systems.

Modern submission infrastructure replaces that chaos with coherence. By centralising control and building visibility across teams, organisations turn frantic document chases into structured collaboration. Every piece of content becomes traceable, every change accountable, and every reviewer confident they are working with the most current version. What emerges is not just efficiency but peace of mind, the knowledge that the system itself supports clarity, rather than undermining it.

*"Submission systems should be as mature
as the devices they represent."*

17.4.2 The Role of Regulatory Information Management

At the heart of this modern infrastructure sits the Regulatory Information Management (RIM) system. Once considered optional, RIM has become foundational for companies operating across multiple geographies. It is far more than a repository of documents, it is a command centre for global submission activity.

A well-implemented RIM links requirements directly to evidence, so no claim is left unsupported. It tracks versions with precision, reducing duplication and guesswork. It manages timelines across global jurisdictions, ensuring that submissions in the U.S., Europe, or Asia are aligned rather than fragmented. Most importantly, it creates visibility: leaders can see where gaps exist, which milestones are at risk, and how well teams are aligned.

When deployed effectively, a RIM becomes the single source of truth. It reduces dependency on individual memory or heroics and builds institutional resilience. Submission stops being a fire drill and becomes a predictable, repeatable process, a capability organisation can rely on again and again.

17.4.3 Modular Authoring - Discipline at Scale

Infrastructure provides the backbone, but how content is created determines its strength. Too often, the same claim, risk statement, or design rationale is re-authored across multiple files. Each iteration drifts slightly, creating inconsistencies that confuse reviewers and erode confidence. This patchwork approach slows review cycles and raises unnecessary questions.

Modular authoring solves this challenge. Instead of rewriting, organisations create structured, reusable building blocks of validated text - clinical claims, risk rationales, device descriptions - each reviewed once but trusted many times. These modules can be assembled, adapted, and localised without starting from scratch.

The result is discipline at scale. Submissions become consistent, coherent, and faster to produce. Reviewers encounter alignment across documents rather than contradictions. Internally, teams spend less time fixing mistakes and more time advancing science. Modular authoring isn't about automation replacing people, it is about freeing people to focus on higher-value work, knowing that the foundation is reliable.

> *"Modular content isn't automation for its own sake, it is discipline made scalable."*

17.4.4 AI and Automation as Enablers

Technology is amplifying this discipline further. Artificial intelligence and automation are beginning to play a meaningful role in submission preparation, not to replace human judgement, but to strengthen it.

AI tools can now flag inconsistent claims across documents, highlight outdated references, and identify unsupported rationales. They can surface contradictions between risk files and clinical narratives that otherwise might slip through under deadline pressure. Some tools even help structure lay summaries for patient-facing content, ensuring clarity across both technical and public audiences.

These are not gimmicks. In a high-volume, high-complexity environment, they are necessities. They do not remove the responsibility to think, they sharpen it. They allow regulatory teams to focus on strategy and substance, while the system takes care of the noise. Used wisely, AI becomes less about novelty and more about enabling rigour at scale.

> *"Use AI not to write for you, but to help you write better."*

17.4.5 Structured Submissions for a Digital World

The regulatory world itself is evolving, and submissions must evolve with it. The era of static PDFs and binders is over. Agencies now expect dossiers to be structured, navigable, hyperlinked, and in many cases, machine-readable. The electronic Common Technical Document (eCTD) is not a future state, rather it is today's reality.

Teams still relying on manual assembly signal unpreparedness, both to regulators and to themselves. By contrast, organisations that embrace structured submissions can respond faster, update easier, and project greater maturity. Structured submissions make the reviewer's job easier, which in turn accelerates approval. They also future-proof the organisation, ensuring it can adapt to regulatory changes without rebuilding processes from the ground up.

17.4.6 Live Traceability – Closing the Loop

Perhaps the most transformative element of operational excellence is traceability. Submissions cannot be treated as static artefacts, frozen in time. They must live, evolve, and stay aligned with the product itself.

In a mature system, a change in a risk file automatically cascades into updates for PMS plans. Adjustments in labelling flow into IFUs and technical documentation without duplication. New clinical data refines claims across dossiers, ensuring that evidence and narrative remain consistent everywhere. This is what live traceability delivers: not just version control but a connected ecosystem where product design, evidence, and documentation move together.

Closing this loop is about more than efficiency. It is about confidence. Confidence that when the auditor arrives, the system is ready. Confidence that when regulators ask a question, the answer is consistent across every geography. And confidence that the organisation is not just compliant in the moment, but resilient over time.

17.4.7 Final Thought – Excellence Engineered

The payoff is significant. Reviewers see consistency instead of contradiction. Teams feel clarity instead of chaos. And organisations build reputations not only for their devices, but for their discipline. Submission quality becomes predictable, not dependent on heroic last-minute efforts.

Submission excellence, then, is not about chasing perfection. It is about building systems that deliver quality under pressure, across jurisdictions, and over time. When infrastructure supports clarity, and clarity builds credibility, the result is trust: from regulators, from internal teams, and from patients waiting for safe and effective technologies.

This is not bureaucracy. It is design. And in a synchronised organisation, it is the engine of competitive advantage.

17.5 Global Collaboration: Strategic Foresight and Regional Respect

In today's MedTech landscape, innovation knows no borders. A device designed in one country is often intended for patients across continents. Yet while technology travels easily, regulatory approval does not. The difference between momentum and delay is rarely technical, it is operational. And it is almost always collaborative.

World-class organisations understand that submission excellence is not a local tactic. It is a global discipline. It requires foresight, respect for regional nuance, and a mindset that treats harmonisation as a strategic lever, not an afterthought.

17.5.1 Replacing the Regulatory Relay Race

Too many companies still treat global submissions like a relay race: file in the U.S., then rework for the EU, and only later adapt for Canada, Australia, or Japan. This sequential mindset creates duplication, introduces inconsistencies, and burdens teams already stretched thin. By the time later markets are addressed, documentation has drifted, alignment has eroded, and the launch momentum is gone.

Synchronised organisations reject this pattern. They architect global launches from the outset, aligning dossier planning with market strategy and engaging cross-regional expertise during development, not after submission.

17.5.2 Global-First Thinking – Build Once, Adapt Intelligently

Global-first thinking asks a simple but powerful question: how do we design for variation instead of being surprised by it? Rather than recreating submissions market by market, companies build a Core Dossier that serves as a harmonised backbone. From there, local adaptations become a matter of refinement, not reinvention.

This approach requires behaviours such as involving regional regulatory experts during clinical and design planning, structuring IFUs with modular language for localisation, and drafting benefit–risk narratives that resonate across diverse regulatory philosophies. It is not about one-size-fits-all, rather it is about one-structure-serves-many.

17.5.3 Strategic Execution in Action

Consider a diagnostics company preparing to launch an AI-powered imaging platform. Instead of working market by market, they synchronised from day one. Clinical studies were designed with endpoints acceptable to both FDA and EU MDR. Claims and risk narratives were aligned in a single source of truth. Submissions were structured, and early engagement with reviewers in the U.S., Europe, and Australia ensured expectations were harmonised.

The result? Submissions filed in three regions within four months of each other, all cleared on the first review cycle. The differentiator was not speed alone. It was intentionality, powered by collaboration.

17.5.4 Embedding Regulatory Intelligence – Seeing Before Reacting

High-performing organisations do not wait for surprises. They embed regulatory intelligence (RI) into development and submission planning, continuously scanning for evolving standards, shifting timelines, and regional nuances.

Strong RI capabilities track emerging standards like IEC 81001 or AI/ ML-specific guidance, monitor agency trends in review behaviour, and identify differences in requirements for cybersecurity, clinical follow-up, or adverse event reporting. This foresight allows teams to pivot early and prepare strategically, turning potential disruption into manageable variation.

17.5.5 Strategic Preview - Designing Beyond Today

Regulatory readiness is not just about passing today's gates, it is also about anticipating tomorrow's. Strategic preview means embedding forward-looking elements into submissions, even before they are mandated. Drafting post-market clinical follow-up (PMCF) plans during initial MDR filings, including cybersecurity summaries aligned with U.S. and EU expectations, or structuring documentation to support future QMSR updates – these are marks of maturity. They demonstrate that an organisation is not only compliant but prepared.

17.5.6 Respecting Regional Nuance - Without Reinventing the Wheel

True global collaboration does not mean erasing differences. It means learning to respect them without losing efficiency. Every region has its own cultural, linguistic, and regulatory flavour, and these differences matter. A submission that feels clear in English may sound bureaucratic or even confusing once translated. A phrase that resonates with FDA reviewers may need to be reframed for EU notified bodies or regulators in Asia. Even when standards are shared, their interpretation is often not.

This is where maturity shows. Leading organisations do not wait until the end to "fix" local gaps, rather they build respect into the process from the start. They engage in-country experts early, not just as translators, but as advisors who understand how regulators think and what language builds trust. They adapt terminology and narrative to local expectations, ensuring that documents feel native rather than transplanted.

The goal is not homogenisation, because uniformity would strip away context and relevance. Instead, the goal is coherence across diversity: a single product identity expressed in many dialects. Harmonisation makes the core consistent, while respect ensures the message lands in every market with credibility.

When this balance is done well, it creates both speed and trust. Regulators see companies that respect their reality, not just tick their boxes. Teams avoid duplication while still delivering submissions that feel tailored. And patients, wherever they are, receive the assurance that the product was designed and explained with their reality in mind.

17.5.7 Final Thought – Beyond Borders, By Design

Global collaboration is no longer optional. It is the price of leadership in MedTech. Organisations that treat submission as a global system, not a series of local hurdles, gain speed, resilience, and trust. They accelerate access, strengthen internal alignment, and build reputations that travel across borders.

> *"Excellence across regions is not luck. It is the result of foresight, discipline, and deliberate connection."*

And in a synchronised model like iQSM, this discipline is not only possible, but also scalable.

17.6 The Mindset Shift – From Afterthought to Advantage

For years, regulatory submissions have been treated as the last mile, the finish line to stagger across rather than a stage to step onto with confidence. This perception has shaped behaviours: teams rush to compile documents, regulatory affairs work in isolation, and executives hold their breath waiting for approval letters. But the Submission Shift shows us a different picture. Submission is not an afterthought, rather it is a mirror. It reflects the discipline of the journey, the strength of alignment, and the culture of the organisation. When approached with that mindset, it transforms from a chore into an advantage, a strategic expression of maturity and credibility.

17.6.1 Submission as a Reflection of Product Maturity

Every submission tells the truth, whether we like it or not. A dossier full of contradictions, inconsistencies, and vague justifications reveals more than weak documentation, it reveals weak systems. It exposes gaps between engineering and regulatory, between claims and evidence, between what was built and how it was explained. Regulators notice. Reviewers are trained to sense when something doesn't add up, and those signals erode trust long before questions are raised.

By contrast, a clear, coherent, and evidence-rich submission demonstrates maturity. It shows that risks were considered early, that testing was not only rigorous but relevant, and that performance claims were carefully aligned with clinical evidence. This kind of submission speaks with authority. It tells reviewers: "This company knows its product. This company understands its responsibility." And that message travels beyond regulators: it resonates with investors, partners, and even patients. Submissions are not paperwork. They are signatures, reflecting the character of the organisation behind the product.

"A submission is more than a report. It is a signature."

17.6.2 Living the Submission Early, Not Writing It Late

The weakest submissions are the ones written in panic, assembled in the weeks before filing, with teams scrambling to stitch together documents that were never designed to fit. In those moments, quality becomes patchwork, evidence feels retrospective, and regulatory affairs are unfairly cast as miracle workers. The result is not just stress, rather it is wasted opportunity.

High-performing organisations avoid this trap because they don't write submissions at the end, they live them from the beginning. Regulatory teams are involved at concept stage, shaping intended use and risk controls before they fossilise. Documentation grows alongside design and clinical planning, so that when the dossier is compiled, it reflects a process already lived, not hastily constructed. Some companies even run mock submissions mid-development, not as busywork, but as pressure-tests for clarity and coherence. The payoff is huge: fewer surprises, faster approvals, and, most importantly, products that carry trust from day one.

17.6.3 Submission Quality as a Shared KPI

In many organisations, submission quality is seen as "Regulatory's job." This narrow view not only isolates regulatory teams but also undercuts the entire product journey. A strong submission is not written in a corner office. It is forged in cross-functional discipline, the alignment of engineering, quality, clinical, marketing, and leadership.

The best companies recognise this and measure submission quality as a shared KPI. They look not just at whether approvals were granted, but at how: Was the dossier coherent? Were claims consistent across CERs, IFUs, and risk files? Did reviewers compliment clarity, or raise avoidable questions? These measures become cultural mirrors. When everyone is accountable for submission quality, it stops being a burden and becomes a craft. Teams take pride in precision. Leaders celebrate clarity, not just compliance. And over time, submission quality becomes not just a metric, but a muscle, one that strengthens the entire organisation.

> *"Submission is not Regulatory's burden*
> *alone, it is a shared achievement."*

17.6.4 Empathy for the Reviewer – Writing with Integrity

At the other end of every submission is a human being. A reviewer, often working under tight timelines and juggling multiple dossiers, who is responsible for ensuring that what reaches the patient is safe and effective. Too often, teams forget this. They bury reviewers under jargon, inconsistent terminology, or endless tables that obfuscate rather than illuminate. The result is frustration, longer review times, and, at worst, mistrust.

Empathy changes everything. Writing with the reviewer in mind means structuring arguments clearly, anticipating questions with evidence, and using consistent language across documents. It means replacing circular justifications with clear logic. It means respecting the reviewer's time and intelligence. Transparency is not a weakness, it is a strength. And when organisations communicate with integrity, regulators respond in kind, with faster decisions, fewer queries, and deeper trust. Empathy is not about being persuasive. It is about being honest in a way that builds confidence.

17.6.5 From Pain to Pride – A Cultural Reframe

For many teams, submission is the least-loved stage of development. It is where enthusiasm wanes, stress rises, and creativity seems to vanish under the weight of checklists. But this view is not inevitable, it is cultural.

When companies reduce submission to paperwork, they drain the energy from their teams. Companies that reframe it as a moment of pride unlock something more powerful: motivation.

When submission is seen as the milestone where everything comes together – design, evidence, risk management, clinical insight – it becomes a celebration of cross-functional achievement. Teams shift from "we need to get this over with" to "we are proud to put this forward." That pride is not decorative, it is contagious. It boosts morale, attracts talent, and reinforces a culture where quality and clarity are valued as much as speed. Submission, in this light, is not the end of innovation. It is its moment of truth.

A clear, well-structured submission communicates control. It tells the reviewer:

> *"We understand our product. We understand*
> *the rules. And we are ready."*

17.6.6 Final Thought – Mindset is the Multiplier

No tool, template, or checklist can replace the mindset with which a submission is approached. Systems enable speed. Templates enforce structure. But it is mindset that multiplies impact. When leaders and teams see submission not as a hurdle, but as a reflection of their best work, everything shifts. Speed improves because alignment is natural. Trust builds because transparency is deliberate. Confidence rises because clarity is embedded.

> *"The Submission Shift, at its core, is not about how you write. It is about how you think."*

And when thinking changes, so does everything else.

Submissions are not the end of the journey. They are the point where innovation becomes visible, where discipline is tested, and where trust is either won or lost. They are not just about approval, they are about advantage. Organisations that embrace this mindset don't just pass regulatory gates, they accelerate through them, ready for scale and success.

As this chapter closes, one truth is clear: submissions reflect systems, and systems reflect culture. Which is why the next step goes deeper than documents. In the following chapter, we explore the leadership and culture that make synchronisation sustainable. Because no system sustains itself. It takes leaders who embody synchrony and cultures that live it daily.

THE CULTURE CATALYST – LEADING FOR ALIGNED IMPACT

In MedTech, systems can be engineered, tools can be deployed, and frameworks can be beautifully designed. But none of it lasts or even functions without the force that breathes life into structure: culture. And the stewards of that culture? Leaders.

This final section of the book is not just the closing chapter, it is the critical culmination of everything we have built so far. If the earlier sections constructed the architecture of synchronisation, this is where we empower the architects. This is the human infrastructure, the mindset, behaviours, and leadership that will determine whether synchronisation becomes an enduring shift or another short-lived initiative.

This section is about legacy. It is about the leaders and teams who will turn this vision into reality, not just once, but repeatedly, reliably, and courageously. It reminds us that the future of MedTech won't be shaped by regulations alone or technology in isolation, it will be shaped by people who choose to lead differently.

Chapter 18 explores what true leadership and transformative culture look like in synchronised MedTech organisations and why they are the final, indispensable ingredients.

The chapter begins by naming a core truth: leadership and culture drive everything. Even the best systems will falter if the people powering them are disengaged or misaligned. Synchronisation is not a process to implement but a spirit to cultivate.

We examine how leaders play multifaceted roles, not just setting direction but championing alignment, mentoring cross-functional teams, and modelling values through action. The strongest leaders don't command synchronisation from above, they embody it visibly and consistently across the organisation.

The chapter then explores culture as the invisible engine behind every success or failure. From psychological safety to continuous learning and collective accountability, it unpacks how values translate into behaviours and how those behaviours shape product quality, team cohesion, and ultimately, patient safety.

We also explore what it takes to lead transformation in the real world: managing resistance, telling compelling stories of change, and building momentum through structure, empathy, and inspiration. Leadership here is both strategic and emotional, a blend of reformer and visionary.

The future-focused lens of this chapter offers a vision of MedTech organisations where synchronisation is no longer a competitive advantage, it is the expected norm. Regulatory bodies are watching not just what we submit, but how we work. Teams that can sustain synchronisation across product lifecycles, technologies, and regions will lead the next era of MedTech excellence.

And finally, we close with a rallying cry: that legacy is not built in boardrooms or spreadsheets. It is built in the quiet moments of collaboration, the brave decisions to break silos, and the everyday choices to lead with purpose. This chapter doesn't just end the book, it hands the baton to the reader.

Without this section, the rest of the book remains theoretical. This section transforms knowledge into action, and action into impact. It ensures the SHIFT isn't something your organisation does, but something it becomes. This is the culture catalyst, the final spark that turns structure into movement, and movement into meaning.

Because the future of MedTech will not be led by frameworks. It will be led by people.

Let that person be you.

SYNCHRONISING FOR SUCCESS – LEADERSHIP, CULTURE, & THE FUTURE OF MEDTECH

S ystems create structure. Processes provide clarity. Technology fuels scale. But without people, without leadership, without culture, none of it lasts.

This is the final frontier of synchronisation: not a new framework or tool, but the human layer beneath them all.

Throughout this book, we have explored how Regulatory, Quality, and Innovation can move in rhythm, aligned across submissions, markets, and product lifecycles. We have seen how strategy becomes speed, how rigour becomes trust, and how convergence reshapes the future of MedTech. But none of that endures unless it is lived, not just documented.

Synchronisation does not thrive because a process says it should. It thrives because someone makes it real. Every great MedTech shift begins with a choice, by a leader who sees across functions, by a culture that rewards alignment, by a team that values transparency more than comfort.

This is not about soft skills. It is about system strength. And the strongest systems are those that can think, adapt, and respond as one.

Leadership is no longer about controlling the centre. It is about synchronising the edges, empowering functions to work together with trust, speed, and purpose. Culture is no longer about corporate posters. It is about what teams prioritise when no one is watching.

As we step into this final chapter, we are not closing the book. We are opening a new one, a future built not only on systems, but on the mindset to sustain them.

Because MedTech is not transformed by process alone. It is transformed by people who lead with integrity, learn with humility, and build with intention.

And it starts now. With us.

18.1 Why Culture and Leadership Drive Everything

At the core of every system, strategy, and success story in MedTech lies one unshakable truth: people make it real.

Processes can be mapped, technologies deployed, and templates perfected, but without leadership to sustain them and culture to nourish them, even the best frameworks eventually lose their power. Synchronisation is not just a technical shift. It is a human one.

18.1.1 The Foundation Beneath the Framework

Building a high-performing system on a weak culture is like planting a tree in poor soil. You may support it with stakes, water it regularly, and protect it from storms, but without nutrients beneath, it will not thrive.

Organisations that treat culture as a secondary concern often struggle to sustain change. They fall into patterns where quality is seen as a barrier, regulatory teams are kept at arm's length, and innovation is pursued in isolation. These behaviours may not be intentional, but they are embedded and dangerous.

Culture is not a slogan. It is the lived experience of how things are done. It determines whether a cross-functional issue is raised early or hidden. Whether a concern is welcomed or avoided. Whether a lesson becomes institutional knowledge or disappears after a project closes.

Leadership and culture together form the soil and sunlight of synchronisation. They cannot be outsourced. They must be cultivated, deliberately and continuously.

18.1.2 Leading with Intention, Not Position

Leadership in synchronised organisations is not about status. It is about stewardship. The most influential leaders are not those with the most authority, but those who model the most alignment.

They ask different questions. Instead of demanding timelines, they ask whether early feedback has been gathered. Instead of reacting to risk, they seek to understand its root. Instead of separating functions, they bring them together to solve problems collectively.

Leadership means being the first to listen, the first to adapt, and the first to model the behaviours others are still learning to adopt.

Synchronisation does not require perfect leaders. It requires present ones: leaders who walk the floor, attend the hard meetings, and treat culture not as a banner, but as an ongoing responsibility.

> *"Synchronisation is not a process you manage. It is a spirit you cultivate."*

18.1.3 Culture as the Air We Breathe

Culture shapes every decision, especially the ones no one sees. It is present in how meetings are run, how trade-offs are made, and how conflict is handled. It lives in what is prioritised, what is ignored, and what is repeated.

In a healthy culture, silence does not signal safety. It signals curiosity. Teams ask questions before drawing conclusions. Individuals raise issues early, not because they are told to, but because they are expected to.

> *"Culture is the air everyone breathes. And in MedTech, that air flows directly into patient safety, product longevity, and reputational resilience."*

This type of culture is not born from luck. It is built through shared values, reinforced by daily behaviours, and protected by leadership.

Examples of strong cultural practice include:

- Encouraging design reviews where Quality leads share risks before solutions
- Normalising post-market reviews as learning opportunities, not blame sessions
- Creating feedback loops where Regulatory, R&D, and Clinical teams share a common language

When culture is aligned, synchronisation does not feel like extra work. It becomes how the work is done.

18.1.4 The Hidden Cost of Misalignment

A fractured culture does not always announce itself. Often, it emerges through slow erosion: trust declines, teams duplicate effort, and decisions are made in isolation. Projects move, but not together. Outcomes are delivered, but not with confidence.

This is where leadership and culture become mission critical. Without them, even the best-designed systems deteriorate under pressure.

Organisations that neglect culture often face:

- Late-stage surprises during regulatory submission
- Frustration between departments due to unspoken misalignment
- Rework that could have been avoided with earlier dialogue

None of these are technical failures. They are cultural ones.

> *"Submission isn't paperwork. It is our passport to global impact. And leadership is about staying in tune with the rhythm of change—not just setting direction and walking away."*

18.1.5 A Human Operating System

Synchronisation is often described in terms of process. But beneath the process is something more powerful: a human operating system based on trust, openness, and accountability.

When leadership and culture are in sync, synchronisation flows. Quality is not a department. It is a mindset. Regulatory is not a hurdle. It is a guide. Innovation is not isolated. It is integrated.

This is not idealism. It is infrastructure. It is what sustains success long after the launch is over, the audit has passed, or the strategy has shifted.

Synchronisation will not last because it is documented. It will last because people choose, day after day, to make it real.

That is what leadership truly means. And that is what culture must protect.

18.2 Synchronised Leadership: Championing the Shift

When we imagine synchronised MedTech organisations, it is easy to focus on streamlined systems and collaborative processes. But behind every elegant operating model is something far more important: leadership that aligns not just actions, but beliefs.

Synchronisation is not a natural state. It is cultivated. And it requires leaders who understand that integration is not a technical task, but a human one, built on clarity, trust, and consistent behaviour.

18.2.1 The Drumbeat of Purpose

In any high-performing organisation, the leader's most important role is to keep purpose visible. Not through grand speeches or annual offsites, but through quiet consistency, the rhythm of meaning that connects daily work to long-term mission.

Synchronised leaders do not simply repeat slogans. They embed purpose into routines. They help teams see that regulatory alignment is not bureaucracy, it is trust. That quality is not a bottleneck but a design choice. That innovation is not freedom from rules, but progress with responsibility.

When leaders make purpose visible, convergence becomes natural. Teams stop asking why alignment matters, because the answer is always there, in the work, in the stories, in the rhythm of leadership.

18.2.2 Leadership in Practice: Visibility Over Volume

Leadership is not measured by how much is said, but by what is consistently done. Synchronised leaders show up early, not just when something breaks. They invite Quality to ideation sessions. They ask Regulatory to help frame early concepts. They do not wait for milestones to involve key voices.

Their presence is not performative. It is participatory. It sends the signal that alignment is not an event, it is a default.

For example, when a CEO attends a cross-functional design review not to lead, but to listen, it changes the room. When the CTO collaborates with the Head of Regulatory on a risk-benefit narrative, the team sees that integration is not a handover, it is a shared responsibility.

These choices speak louder than strategy decks. They show that synchronisation is not something expected from others, but something modelled at the top.

18.2.3 Empowering Through Trust and Structure

Synchronised organisations do not rely on individual heroics. They rely on distributed clarity. Empowerment is not the absence of oversight. It is the presence of trust within aligned systems.

> *"You don't lose trust by admitting a*
> *problem. You lose trust by hiding it or by*
> *ignoring the person who spoke up."*

This means setting clear boundaries, shared language, and mutual respect across roles. Engineers are trusted to raise safety concerns without fear. Regulatory teams are empowered to contribute upstream, not just validate downstream. Quality professionals are included as co-designers, not compliance monitors.

One powerful approach is cross-functional mentoring. Pairing RA and R&D leads, rotating quality specialists through clinical teams, or involving commercial leaders in usability reviews. These practices build not only understanding but mutual investment.

Empowered teams are more than productive. They are proactive. They challenge assumptions early, prevent delays later, and respond to change without panic.

18.2.4 The Quiet Influence That Lasts

Effective leadership is not always loud. It is consistent. It replaces escalation with alignment. It favours listening over assumption. It trades ego for clarity.

These habits might go unnoticed in the moment. But they build trust, defuse tension, and make synchronisation feel natural rather than forced.

Leadership in this context is not about being the hero. It is about being the harmoniser, the one who ensures that the right voices are present, that decisions reflect shared goals, and that purpose is not only stated, but felt.

When leaders lead this way (visibly, quietly, with discipline and heart) synchronisation becomes more than a model. It becomes the way things are done.

And in doing so, they do not just build better products. They build stronger organisations, ones that last, adapt, and lead with confidence.

18.3 Culture in Action: From Values to Daily Rhythms

Culture is not what is written on office walls or websites. It is what people do when no one is watching. It is reflected in the conversations teams avoid, the decisions they defend, and the habits they reinforce under pressure. In MedTech, where the stakes involve not just performance but patient lives, culture is not a backdrop. It is the engine.

Synchronisation can only thrive if the culture enables it. For organisations to move as one, culture must reward openness, reinforce shared purpose, and support intelligent risk-taking. It is not abstract. It is operational. It shows up in daily rituals, in how feedback flows, and in how people respond when things go wrong.

18.3.1 Making the Invisible Visible

The first sign of a healthy culture is that visibility is not feared. It is expected. People surface problems early because they trust the system will respond with action, not blame.

When engineers raise usability concerns or Regulatory challenges a product claim, the question is not whether they are slowing progress. The question is whether they are helping avoid greater risk down the line.

Leaders must model and protect this openness. That means welcoming challenge, asking honest questions, and rewarding truth even when it complicates plans.

Simple, repeatable practices can strengthen this transparency:

- Building time into reviews for "what concerns us"
- Highlighting cross-functional contributions in team briefings
- Creating informal channels to share lessons, not just successes

These practices signal that visibility is not just safe, it is essential. They create a culture where silence is not mistaken for progress.

18.3.2 Learning Loops That Strengthen Over Time

In fast-moving, high-regulation environments, learning cannot be reactive. It must be continuous.

Real learning happens when insights from complaints, audits, or field feedback are brought upstream and when evidence is not just collected but applied.

> *"Culture isn't built on knowing everything.*
> *It is built on wanting to know more."*

Organisations with strong learning cultures:

- Use post-market signals to refine risk files and test plans
- Turn adverse event reviews into team-wide design discussions
- Include Regulatory in usability test reviews, not just reports

One organisation introduced a monthly reflection forum where Quality, R&D, and RA shared one missed opportunity and one insight applied. Over time, these short, shared learnings created a sense of maturity and self-correction.

When feedback loops are active and respected, improvement becomes part of the operating rhythm, not a special project.

18.3.3 Aligning Incentives with Purpose

Synchronisation falters when teams are rewarded in isolation. If R&D is measured only on speed and RA only on compliance, natural tensions grow.

Culture becomes fractured when what is celebrated in one part of the organisation contradicts what is required elsewhere.

Organisations that sustain synchronisation build shared incentives:

- Cross-functional goals in project scorecards
- Joint ownership of key milestones
- Celebrating integrated problem-solving, not functional heroics

It is not about removing accountability. It is about aligning it. When success is defined together, collaboration becomes the path of least resistance.

18.3.4 Adapting with Resilience and Trust

No organisation is static. New regulations emerge. Teams restructure. Markets shift. In these moments, culture is tested.

"Resilience isn't about avoiding disruption.
It is about adapting faster, together."

Some teams pause, wait for direction, or default to their silos. Others engage, realign, and adapt together. The difference is rarely structural. It is cultural.

Adaptable cultures create space to ask:

- What can we learn from this disruption?
- Which functions need to re-coordinate?
- Where does clarity need reinforcing?

One diagnostics company facing a critical materials shortage created a cross-functional rapid response group. The focus was not just resolving the issue but learning from it. They updated their sourcing strategy, engaged regulators proactively, and improved traceability protocols as a result.

This was not a one-time response. It reflected cultural muscle built over time, a culture that sees change not as a threat but as a trigger for synchronisation.

18.3.5 Culture as a Strategic Asset

Culture cannot be downloaded or declared. It must be demonstrated, in moments of tension, in how teams speak when they disagree, in whether the system invites questions or suppresses them.

A culture that supports synchronisation does not happen by accident. It is designed through intention, reinforced through daily behaviour, and protected by leaders at every level.

This type of culture enables:

- Early risk identification
- Faster, more aligned decision-making

- Greater resilience under regulatory pressure
- Teams that don't just work well, but work well together

It is easy to overlook culture because it is hard to measure. But in the long run, it is the most accurate predictor of whether synchronisation will last.

When culture reflects shared values and collective purpose, alignment is not forced. It flows.

18.4 Leading Transformation: The Reformer and the Visionary

True transformation in MedTech does not happen by accident. It requires a specific blend of mindsets: the grounded clarity of the reformer and the expansive vision of the dreamer. One focuses on structure, the other on possibility. One sees the detail that must be fixed, the other sees the future that must be built.

This balance is what drives synchronisation forward. Without reform, vision fades into abstraction. Without vision, reform becomes routine. Together, they shape the systems, culture, and strategy needed to make synchronisation real and sustainable.

> *"Be the reformer who builds patiently. Be the visionary who dares greatly. The future belongs to those who can be both."*

18.4.1 Treating Change as a Strategic Programme

Many organisations treat cultural and operational change as a communications exercise. They focus on alignment sessions, vision statements, and posters. But synchronisation requires more. It must be treated like any other critical programme: with sponsorship, structure, and discipline.

Successful transformation efforts share four elements:

1. Sponsorship

Change only becomes real when it is led from the top. Senior leaders must not only approve initiatives, they must also participate in them. A transformation cannot live on slide decks alone. It lives in the choices leaders make every week.

2. Roadmap

Clarity builds momentum. A phased, realistic roadmap makes synchronisation feel actionable. For example:

- Phase one may focus on building shared language and early collaboration rituals
- Phase two might introduce integrated decision forums and cross-functional metrics
- Phase three could align governance, incentives, and post-market feedback systems

3. Metrics

Transformation is hard to measure, which is why most initiatives lose energy. But thoughtful signals help:

- Proportion of new projects with early RA and Quality input
- Reduction in late-stage rework
- Engagement scores from cross-functional teams
- Number of lessons applied across multiple product lines

4. Rituals

The most enduring change happens not in formal events, but in rhythms. Teams that succeed embed transformation into the way they meet, plan, and reflect:

- Monthly synchronisation reviews
- Cross-functional retrospectives
- Stories of integration shared in all-hands briefings

Treat synchronisation like a programme, not a campaign, and it will become embedded in the way the organisation thinks and moves.

18.4.2 Building the Narrative That Moves People

Facts inform. Stories move. Transformation only takes root when people see themselves in the change. That means telling the story of synchronisation in a way that connects emotionally, not just logically.

A compelling transformation narrative often follows three movements:

1. Where we have been: Acknowledge the friction and fragmentation. Describe what has held the organisation back, not to assign blame, but to show awareness.
2. What is changing: Explain how teams are shifting from isolated effort to shared rhythm. Clarify how systems, incentives, and mindsets are being restructured for better outcomes.
3. Where we are going: Describe a future that feels tangible. Not only the benefits to the company, but the impact on teams, patients, and trust.

At one global firm, a senior leader opened each town hall with a short story of synchronisation in action. One told how a design team avoided a three-month delay by aligning their risk documentation early. Another highlighted how RA flagged a subtle claim inconsistency that would have caused rework across three regions. These were not success stories. They were synchronisation stories. And over time, they reshaped the organisation's sense of identity.

"If you want people to build the future, show them the bricks — but more importantly, show them the castle."

18.4.3 Responding to Resistance with Curiosity

Resistance to change is natural. It is often misread as defiance when it is really uncertainty. Leaders who succeed in transformation treat resistance as insight, not opposition.

Common reframes include:

- This will slow us down → This prevents rework later and helps us move with confidence
- This is adding complexity → This creates structure where the risks are highest
- We have never done it this way → That may be the clearest reason it is time to explore

Reframing is not about pushing harder. It is about connecting. One organisation introduced integrated RA and Quality checkpoints into early development. A senior engineer initially opposed the change. Instead of enforcing it, leadership invited a pilot phase. After seeing how much time was saved in risk review, that same engineer became a vocal advocate.

Resistance is not a sign of failure. It is a chance to build trust, show value, and grow alignment through real experience.

18.4.4 Balancing Structure and Inspiration

The reformer brings the systems. The visionary brings the spark. To lead lasting change, organisations need both. The structure of synchronisation must be visible and operational. The spirit of synchronisation must be felt and shared.

This blend looks like:

- Strategic plans tied to compelling purpose
- Governance forums that begin with user impact
- Metrics that include collaboration and trust, not just velocity
- Leaders who coach, co-create, and share credit

Transformation in MedTech is neither fast nor easy. It takes intention, resilience, and belief. But when leaders embrace their dual roles – builder and believer, planner and storyteller – they create the conditions where synchronisation is not just a shift, but a legacy.

18.5 The Future of MedTech: Sustaining Synchronised Success

The future of MedTech will not be shaped by breakthroughs alone. It will be defined by the organisations that can deliver those breakthroughs consistently, responsibly and at scale. In this future, synchronisation is no longer a novel advantage. It is the expected standard.

Companies that align their regulatory, quality, and innovation functions will move faster with confidence. They will adapt to complexity without chaos. And they will earn the trust of both regulators and patients, not through promises, but through practice.

Synchronisation is not a trend. It is the foundation of sustained excellence.

18.5.1 Agility with Integrity: Moving Fast and Right

Speed has always been prized in product development. But in MedTech, speed without integrity leads to risk. Integrity without speed results in missed opportunities. Synchronised organisations do not choose between the two. They combine them.

This new agility is not reactive. It is anticipatory. It is built on upstream alignment and continuous awareness. Teams engage early, reduce rework, and solve problems before they scale.

Key markers of agile synchronisation include:

- Upfront collaboration between engineering, regulatory, and quality teams to define shared assumptions and constraints
- Rapid integration of post-market feedback into design cycles, risk files, and labelling
- Parallel preparation for global submissions, reducing regional lag

One global company introduced early cross-functional design reviews that replaced formal gatekeeping with dialogue. The result was a 30 percent reduction in development timelines, improved clarity for regulatory bodies, and fewer late-stage surprises.

Synchronisation does not slow you down. It accelerates what matters.

"The future belongs to those who understand that agility isn't speed alone — it is readiness in motion."

18.5.2 Earning Trust in a Transparent World

The MedTech landscape is shifting. Regulators are raising expectations, not just for the product but for the organisation behind it. Increasingly, questions are not limited to "Is this safe?" but extend to "Is this company operating with maturity and foresight?"

This evolution is already visible:

- The FDA's Case for Quality encourages life cycle thinking rather than snapshot compliance
- Global bodies such as IMDRF are standardising documentation expectations and promoting transparency
- Agencies are actively evaluating whether a company can evolve safely as science and technology advance

Organisations that operate with synchronisation built into their DNA are already aligned with these rising expectations. Their systems are transparent, their files are traceable, and their culture is resilient.

This is not about polishing documentation for inspection. It is about building practices that hold up under scrutiny, because they were designed for it.

Synchronisation makes compliance easier because it is built into how teams think and collaborate, not tacked on at the end.

18.5.3 Measuring What Matters: From Metrics to Meaning

Success in the future of MedTech will not be defined solely by time to market or the number of approvals won. It will be measured by trust sustained and outcomes improved.

Synchronised organisations use metrics that reflect both performance and impact. They ask:

- Are we learning from real-world evidence and applying it upstream?
- Are our changes improving safety, usability and clinician confidence?
- Are we reducing unnecessary complexity without sacrificing rigour?

A leading MedTech company reengineered one of its catheter systems after usability data revealed subtle but important limitations. Rather than isolating the fix, they applied the learning across multiple product lines. Complaint rates dropped, regulator confidence rose, and customer satisfaction improved. The changes were measurable, but the mindset behind them was what mattered most.

Synchronisation is not just about being faster or more efficient. It is about being more trusted and more effective, consistently.

18.5.4 Building the Infrastructure for Future-Readiness

To thrive in the years ahead, MedTech companies must shift from project-based success to system-level resilience. That means embedding synchronisation into how teams operate, how systems connect, and how decisions are made.

This future-focused infrastructure includes:

- Shared repositories of reusable evidence that feed submissions across markets
- Integrated governance models that bring RA, QA, and R&D into one decision cycle
- Digital tools that track traceability, document history, and post-market signals in real time

More importantly, it includes a culture where learning is ongoing, feedback is welcomed, and change is not feared but anticipated.

In this future, synchronisation is not the exception. It is the expectation. Organisations that master it will move more confidently, respond more swiftly, and build stronger relationships across the entire healthcare ecosystem.

18.5.5 The Human Impact: What Success Really Means

At the centre of all this strategy, structure, and speed is the patient. The real test of synchronisation is whether the right solution reaches them faster, safer, and with fewer barriers.

This is not theoretical. Every delay avoided, every misunderstanding resolved, and every risk mitigated through synchronised practice contributes directly to human health.

- When post-market data flows cleanly into early design
- When cross-functional teams anticipate submission needs before drafting begins
- When usability concerns are raised, not hidden

These are not procedural wins. They are outcomes that ripple into clinical care, caregiver trust, and global access.

The companies that will shape the next chapter of MedTech are not those who master one market or product. They are those who master how to work together, across functions, regions, and roles with clarity and purpose.

Synchronisation does not just help companies succeed. It helps healthcare deliver.

18.6 The Final Rallying Cry: Legacy Through Synchronisation

As this journey draws to a close, we are not standing at an endpoint but at a point of inflection. Synchronisation is no longer an idea or a framework. It is now a choice. A practice. A mindset that can shape not just how we operate, but why we operate.

This is the moment to decide what kind of MedTech future we will build. One that protects boundaries or one that bridges them. One that delivers compliance or one that delivers impact. The difference is not tools. It is intention.

18.6.1 Small Acts That Start Movements

Legacy does not begin with bold declarations. It begins with a single choice to act differently. To speak up, to invite collaboration earlier, to simplify one document that others rely on.

You do not need a new role, a bigger team, or a revised strategy. You need five minutes of clarity and a willingness to move first.

A few ways to start:

- Invite your Quality or Regulatory counterpart into a design discussion before it is formal
- Share a synchronisation success at your next team meeting and highlight how the win happened
- Map a post-market insight back to its source document and ask what could be improved
- Pair a junior engineer with a Regulatory mentor for a single project
- Add a standing item to team meetings focused on cross-functional opportunities, not just updates

These are not grand initiatives. They are cultural signals. They say, "this is how we work now." They become stories others follow.

And when enough people choose to lead in this way, synchronisation stops being an effort and becomes an instinct.

> *"Transformation doesn't crash through the door — it grows in the quiet courage to ask, 'What if we did this together?'"*

18.6.2 Defining the Endgame: What Success Really Looks Like

The real finish line is not a cleared submission or a successful audit. It is the impact of that work in the world. The patient who receives the device six months earlier. The clinician who trusts the product because every instruction made sense. The regulator who sees not only compliance, but competence.

Synchronisation enables this. Not by eliminating process, but by refining it. Not by adding complexity, but by clarifying roles and timing. Not by erasing accountability, but by distributing it.

In this model:

- Quality is designed in, not checked later
- Regulatory is strategic, not reactive
- Innovation is integrated, not isolated

> *"Tomorrow's fastest approvals will go to*
> *today's most aligned organisations."*

This is not a dream. It is already happening in forward-thinking organisations that choose to work differently. Where the rhythm of collaboration becomes the operating system. Where no one asks, "who owns this?" Because the answer is always, "we all do."

18.6.3 The Legacy That Lasts

Legacy is not about being remembered. It is about what continues because of you. The culture you shape, the conversations you start, the decisions you influence – all these echo beyond your tenure.

Imagine:

- A young engineer joining your team and believing that cross-functional alignment is normal, because you made it so
- A regulatory professional raising a risk early and feeling supported, because you built that environment
- A patient benefiting from safer, more accessible technology, never knowing that your choice to synchronise made the difference

These are not distant effects. They are the reason we do this work.

Legacy is not about status. It is about stewardship. About leaving something stronger, clearer, and more human than how we found it.

> *"Legacy isn't about being remembered. It is about what continues because of you."*

18.6.4 The Beginning of Your Chapter

This is not just a call to reflect. It is a call to lead. Synchronisation is not sustained by systems, but it is sustained by people. Ask yourself and your teams bold, practical questions:

- Are you actively modelling the values and behaviours you want to see in your organisation?
- Where does your leadership style fall on the spectrum between control, coordination, and true empowerment?
- Is your culture helping synchronisation flourish or silently reinforcing silos?
- What will it take to move from one-time transformation to sustained cultural rhythm?
- What legacy do you want to leave as a leader, as a team, and as an organisation?

So, as we close this book, we leave you not with an answer, but with a question. When someone looks back at the work you led, will they see a product that simply launched? Or will they see a product that learned, adapted, and stood the test of time?

Will they see a culture that worked around each other? Or a culture that worked with each other?

"Great MedTech doesn't just reach the market faster — it stays safer, longer."

The synchronisation shift is not something to admire. It is something to build. Every day, in every meeting, in every interaction. This is not the end of the story. It is the first line of yours.

Let us synchronise.
Let us serve.
Let us lead.

The future of MedTech is not something we wait for. It is something we create: together.

www.ingramcontent.com/pod-product-compliance
Lightning Source LLC
Chambersburg PA
CBHW030449210326
41597CB00013B/598

9 781805 418467